Community Nutrition in Action
An Entrepreneurial Approach

SIXTH EDITION

Marie A. Boyle, PhD, RD
College of Saint Elizabeth

David H. Holben, PhD, RD
Ohio University

Prepared by

Patricia Beffa-Negrini, PhD, RD
University of Massachusetts, Amherst

Nicole Geurin, RD

Denine Stracker, RD

Amanda Sylvie, RD

WADSWORTH
CENGAGE Learning·

Australia • Brazil • Japan • Korea • Mexico • Singapore • Spain • United Kingdom • United States

For product information and technology assistance, contact us at **Cengage Learning Customer & Sales Support, 1-800-354-9706**

For permission to use material from this text or product, submit all requests online at **www.cengage.com/permissions** Further permissions questions can be emailed to **permissionrequest@cengage.com**

ISBN-13: 978-1-111-99006-0
ISBN-10: 1-111-99006-9

Wadsworth
20 Davis Drive
Belmont, CA 94002-3098
USA

Cengage Learning is a leading provider of customized learning solutions with office locations around the globe including Singapore, the United Kingdom, Australia, Mexico, Brazil, and Japan. Locate your local office at: **www.cengage.com/global**

Cengage Learning products are represented in Canada by Nelson Education, Ltd.

To learn more about Wadsworth, visit **www.cengage.com/wadsworth**

Purchase any of our products at your local college store or at our preferred online store **www.cengagebrain.com**

Printed in the United States of America
1 2 3 4 5   16 15 14 13 12

☆ Acknowledgements ☆

I would like to thank
Marie A. Boyle, PhD, RD
for her guidance on this project.

I wish to take this opportunity to thank
Nancy L. Cohen, PhD, RD, LDN
my advisor, mentor, and friend
who was the instructor for my first
course in Community Nutrition

and

my students and co-authors
who inspire me daily.

☆ Dedication ☆

To

David B. Beffa-Negrini

my companion, confidant, and cheerleader.

— Patsy Beffa-Negrini, March 29, 2012

☆ Table of Contents ☆

☆ Project Introduction ☆

This second edition of the *Community Needs Assessment Workbook* is designed to help nutrition and allied health students organize and conduct a community nutrition needs assessment, particularly of a city, county, or state. This workbook follows the "Steps in Conducting a Community Needs Assessment" from Boyle and Holben's *Community Nutrition in Action: An Entrepreneurial Approach, Sixth Edition.*

In community nutrition, a specific geographic or social community becomes the "patient." Any plan for nutrition intervention services should be based on an assessed need, with program objectives and methods outlined. Using the *Community Needs Assessment Workbook*, you will be guided through the basic steps of assessing the nutrition education needs of a population. In some cases, your instructor may ask you to use this community needs assessment later in the semester to plan a nutrition intervention in the community of interest.

Learning Goals and Objectives

Students completing this *Community Needs Assessment Workbook* should be able to assess a community's nutrition problem and needs using principles of epidemiology as appropriate.

By the end of this community needs assessment you will have:

- written a statement that defines a community's nutrition problem.

- set goals and measurable objectives for collecting data about a population.

- collected objective and/or subjective data about a community or individuals who represent a target population.

- identified, evaluated, and prioritized a community's nutrition problems and needs.

- shared the findings of a community nutrition assessment.

- described a plan of action for what should be done next to begin working to solve an identified nutritional problem.

Overview

As you begin the activities in this workbook, you will first brainstorm to identify a population that you would like to investigate. Nutrition assessments like those described in the *Community Nutrition in Action: An Entrepreneurial Approach* textbook may be quite extensive and take many months to conduct. For this project, which will be completed in a few short weeks, the most important thing is to stay focused.

Once you have decided on a group or population of focus, then you will

- set the parameters of the assessment by stating the purpose of the assessment, defining the target population you will assess, setting goals and measurable objectives for the needs assessment, and listing the types of data you need to collect.

- collect data about the broader community (including community, environmental, and socioeconomic characteristics), background conditions (national policies and cultural characteristics), and your particular target population.

- analyze and interpret the data.

- share the findings of the community needs assessment with stakeholders, agencies, community organizations, or others who may benefit from your research.

Based on the community needs assessment, many important issues will be identified. The next step will be to work to prioritize your findings and choose an action plan for what you would like to do next to improve the nutritional status of those in your target population.

☆ Final Community Needs Assessment Report ☆
Note: As you work through this workbook, pay particular attention to gray boxes that are similar to this one. Information you include in these boxes will be used to write your final community needs assessment report.

☆ Step 1: Identify a Population and Define a General Nutrition Problem ☆

Overview

In your community, are there nutritional problems that are not being addressed? Surely, you have answered this question in the affirmative. Are there educational programs or other types of services available to combat these problems? That question may be a bit more difficult to answer without doing some research.

The community needs assessment is a process that community nutrition professionals go through to recognize a nutrition problem, identify community resources to solve the nutrition problem, and prioritize strategies to help improve the nutrition issue in the identified population.

Practicing community nutritionists typically work to identify nutrition problems in clients or patients at their place of employment or in the surrounding community. As a student of community nutrition, you should look to find a population that you would like to assess based on your personal interest, while at the same time considering your personal transportation issues, availability of or need for appropriate supervision, and professional liability.

Your instructor may ask you to design a nutrition education program later in the semester based on the outcome(s) of this assessment project. At this point, depending on your education and professional credentials, you may or may not be able to implement a nutrition education project in a population that has serious health issues. But you might be able to help provide a nutrition education program for people who are generally healthy such as fellow college students, patrons at a local food bank or homeless shelter, or a neighborhood after-school program.

During this first attempt at completing a community needs assessment, choosing a population to investigate may be challenging. During the next activity, you'll use the "Identify a Population and Define a Nutrition Problem" worksheet, which should help you pinpoint a group to assess.

Activity

Use the "Identify a Population and Define a Nutrition Problem" worksheet on the next page to

1. identify a population or brainstorm populations that you would like to work with this semester.
2. choose two populations and list potential nutrition issues that could be important to these populations.
3. list names and contact information for people in the community you can contact regarding this project.
4. list team members, if your instructor asks you to work on this assessment in a small group.
5. determine a target population.
6. conduct library research to find background information about an identified population and nutrition problem.
7. write a statement of the nutritional problem.[1]

---

[1] Step 7 will be used as part of your final Community Needs Assessment Report.

4

☆ Worksheet for Step 1: Identify a Population ☆
and Define a Nutritional Problem

Complete the following individually:

1. Think about the population(s) that you work with or people who live in your surrounding community. List potential groups who might have nutritional problems and need nutrition education. Remember to stay focused. For example, hungry or food insecure populations may include two primary groups: homeless and working poor. Focusing on the working poor can lead to a target population of families accessing a specific local food pantry.

_____   _____

_____   _____

_____   _____

_____   _____

_____   _____

2. From the list above, choose the top two groups you would most like to work with. Below, hypothesize about potential nutrition problems that would likely need nutrition education in each group. For example, if you identified middle-aged women, they may have health issues such as obesity or hypertension or they may need help with budgeting or preparing foods safely to prevent foodborne illness. Likewise, parents who send lunches with their preschool-aged children may need ideas for healthy and quick-to-prepare meals that are nutritionally balanced.

Group 1: _____

Potential nutrition problems: _____

_____

_____

_____

_____

_____

Group 2: _____

Potential nutrition problems: _____

_____

_____

_____

_____

_____

3. Use the telephone book, Internet, list of employees at your workplace, etc. to find and list potential organizations and/or people to contact as you work on this community needs assessment project (and subsequent nutrition education project if assigned.) Also think of other contacts such as people in the community who work with the population of interest on a daily basis. Their perspectives will give you a broader vision of the true nutritional issues and underlying problems. Write a list of potential contacts or stakeholders here:

Contact	Phone or e-mail address	Date contacted

4. If this is a group project, list your team members (and contact information if needed):

_____   _____

_____   _____

_____   _____

_____   _____

6

5. Identify the population you will assess during this community needs assessment and the nutrition problem of focus. (You may need to meet with your team members to complete this step). This problem may come from the list above, or it may come to you after contacting community contacts during step 3.

6. Use your campus library, online library, or Google Scholar to find at least 5 research articles published within the last 10 years about your chosen population and the nutrition problem of focus. For example, if you would like to assess food safety education needs of students at a local elementary school, find published research on food safety education for K-12 students.

   You will use the research articles to write a "Statement of the Nutritional Problem" for the introduction of your final nutrition assessment report.

   Write the citations for the research articles here:

   1.

   2

   3.

   4.

   5.

7. Write a draft "Statement of the Nutritional Problem." Be sure to include information from the references above to support your statement. This statement will be used as the introduction to your community needs assessment report and the references will be cited at the end of the report. (Examples of these types of statements can be found in Table 2-1 on page 43 in your textbook.)

☆ Statement of the Nutritional Problem ☆

8

☆ Step 2: Set the Parameters of the Assessment ☆

Overview

At this point, you should have identified a community group that you would like to assess and a nutrition problem that is of interest to that group. As with any journey, it's important to plan ahead. By setting the parameters for your community needs assessment, you will have a road map to guide the needs assessment.

Writing the purpose of the assessment and branching off from there to the broader goals and specific objectives will help guide you, keep you on track, and prevent gaps in the assessment (omissions of critical information). This step will also help you identify the types of data to collect about the nutritional status of the target population.

Activity

Use the "Community Needs Assessment Parameters" worksheet on the next page to

1. briefly describe the community you are assessing.

2. state the overall purpose of the assessment.

3. describe the community's smaller target population, which will be the focus of the community needs assessment.

4. write the broad goals of the assessment.

5. use active verbs to write the measurable objectives for carrying out the assessment.

6. list the types of data you need to collect about the target population.

✮ Worksheet for Step 2: Community Needs Assessment Parameters ✮

Work either individually or, if directed by your instructor, with your project group to fill in the following.

1. Briefly, in a short sentence or statement, describe the broad community or geographical location where you will conduct your community needs assessment. This statement will become the "Definition of the Community" for your final report.

_____

_____

✮ Definition of the Community ✮

2. Why are you conducting this particular assessment? What is its purpose?

_____

_____

_____

✮ Purpose of the Assessment ✮

3. It is important for your community needs assessment to have a focus. To assess the nutritional status of an entire community is a huge task and would take more time than you have during this assignment. It is important to refine and focus your efforts on a target population. Describe your target population here:

_____

_____

10

4. The goals and objectives for the community needs assessment will help you to focus on the types of data you need to collect about the target population. Write at least 2 goals for the community needs assessment.

Keep in mind that goals are broad and state what is to be accomplished during the nutritional assessment. For example, "The nutrition knowledge, attitudes, and skills of parents of preschool children at ABC Child Care Center will be determined." Other examples can be found on page 43 in your textbook.

Goal A: _____

_____

Goal B: _____

_____

Goal C: _____

_____

5. The next step is critical and will help you determine the types of data you will need to collect in the next section. In this step you will write your objectives. Keep in mind that your goals will be broad and general while the objectives are specific and include active verbs such as *assess*, *identify*, *list*, *increase*, *determine*, etc. The objectives must be measurable and clearly written, and each one should focus on a single idea or purpose. (See examples for writing measurable objectives in Appendix C of this workbook.)

Begin by re-writing each of your goals in the space provided, and then write at least 2 measurable objectives for each of the goals.

Goal A: _____

_____

• Objective 1: _____

_____

• Objective 2: _____

_____

- Objective 3: _____

_____

Goal B: _____

_____

- Objective 1: _____

_____

- Objective 2: _____

_____

- Objective 3: _____

_____

Goal C: _____

_____

- Objective 1: _____

_____

- Objective 2: _____

_____

- Objective 3: _____

_____

Finally, choose 2 of these goals and the associated objectives and re-write them in the gray box. You will use these in the final step of this section to help you specify the types of data that you will need to collect about the target population. This part will also be included in the final report.

12

☆ Set the Goals and Objectives for the Needs Assessment ☆

Goal 1: _____

_____

- Objective 1: _____

_____

- Objective 2: _____

_____

- Objective 3: _____

_____

Goal 2: _____

_____

- Objective 1: _____

_____

- Objective 2: _____

_____

- Objective 3: _____

_____

---

6. Look again at your goals and objectives. You will need to collect data based on the purpose, goals, and objectives of the community needs assessment. You may have to do some library research to find the data specified in your goals and objectives. (Optional: Another way to collect data would be to work directly with representative members of the target population by conducting a survey or focus group.)

The data you collect will be used to identify people in the target population who are at risk, describe the nutritional problem in more detail, identify nutrition needs that are currently not being met, or show you where services are being duplicated and may not be necessary. Keep in mind that what you write as you complete this step is just a first draft and may need to be modified as you begin collecting data.

The types of data you will need to collect can fall into these broad categories: background conditions, community conditions, and target population (living, working, and social conditions; and individual lifestyles factors).

Background Conditions. What are the national policies or cultural characteristics that are important to the target population? List sources where you might be able to find data about these issues. International and government resources are particularly helpful here. You might also look at the research articles you found in Step 1 for sources.

Issue 1: _____

Potential data sources: _____

_____

Issue 2: _____

Potential data sources: _____

_____

Issue 3: _____

Potential data sources: _____

_____

Issue 4: _____

Potential data sources: _____

_____

Issue 5: _____

Potential data sources: _____

_____

14

Community Conditions. From the list below, check which ones will be important for you to assess in order to thoroughly assess your target population. Also identify whether the data you collect will be qualitative (from key informants or stakeholders) and/or quantitative (databases, vital statistics, published research studies, etc.).

Condition	Check if important to the community needs assessment	Check if data collected will be qualitative	Check if data collected will be quantitative
Community Characteristics			
Community organizational power and structure			
Demographic data and trends			
Existing community services and programs			
Community opinion			
Environmental Characteristics			
Food systems			
Health systems			
Housing			
Recreation			
Transportation systems			
Water supply			
Socioeconomic Characteristics			
Sociocultural data and trends			
Economic data and trends			

Target Population Data. To collect data about your target population, you can use existing data that has already been collected or, in the case that important data do not exist, collect new data using such methods as surveys, focus groups, interviews, or other direct assessment techniques.

What questions do you have about your target population's nutritional problems? What do you need to know to complete a through community needs assessment?	How would you collect this data?	
	Collect using existing data (specify potential source[s])	Collect new data (specify method to be used)
Individual Lifestyle Factors and Food Choices		
Food Supply or Food Availability		
Health Beliefs and Practices		
Living and Working Conditions		

☆ Step 3: Collect Data About the Community ☆

Overview

You have goals and objectives for your assessment. You have questions about the health and nutritional status of your target population and potential ways to answer those questions using existing data or by collecting new data using methods such as surveys, focus groups, or interviews. Now it's time to put on your researcher's hat and begin collecting data about the *background conditions*, *community (community conditions, environmental conditions, and socioeconomic conditions)*, and *target population*.

During this step you will collect already-existing objective information about the nutrition needs of the community. You will also interview at least one key informant to collect subjective data. In addition, you will think of other types of data you would like to collect if you had several weeks to do so (see Chapter 3 in *Community Nutrition in Action*, "Types of Data to Collect About the Target Population"); this could become a project later in the course if assigned by your instructor.

Collect Objective Data: Gather information from your college or university library, your local library, and/or the Internet, as well as by calling town offices and agency staff in the community. Include demographic and socioeconomic information, health and vital statistics, local health resources, cultural factors, and availability of housing, food, community and school nutrition programs, etc.

These online resources (plus those listed in Appendix B of this workbook) may be useful:

* FEDstats.gov: http://FEDstats.gov

* Centers for Disease Control and Prevention: http://www.cdc.gov/

* County Health Rankings: http://www.countyhealthrankings.org/

* Community Nutrition Mapping Project:
  http://www.ars.usda.gov/Services/docs.htm?docid=15656

* Your Food Environment Atlas: http://ers.usda.gov/foodatlas/

* Youth Risk Behavior Surveillance System (YRBSS):
  http://www.cdc.gov/healthyyouth/yrbs/index.htm

Collect Subjective Data: Speak with relevant community members to gather subjective information. You may need one or more site visits to the community; plan to call community members as well. A local telephone book can be helpful.

Show all of your data and its source. Following are examples (for data from a report, personal conversation, and website, respectively):

County X has a 40% mortality rate from cancer. Source: MA Dept PH Report on Cancer, 1997.

School X has high rate of teen births. Source: Mrs. Jones, school nurse, personal conversation March 22, 2012.

In Alabama in 2009, 26.3% of high school students surveyed did not drink 100% fruit juices during the 7 days before the YRBS survey. Source: Centers for Disease Control and Prevention.

(2009) CDC Youth Online: High School YRBS 2009 Results. Retrieved on March 29, 2012 from http://apps.nccd.cdc.gov/youthonline/App/Results.aspx?TT=&OUT=&SID=HS&QID=H72&LID=&YID=&LID2=&YID2=&COL=&ROW1=&ROW2=&HT=&LCT=&FS=&FR=&FG=&FSL=&FRL=&FGL=&PV=&TST=&C1=&C2=&QP=G&DP=&VA=CI&CS=Y&SYID=&EYID=&SC=&SO=.

Activity

Use the "Data Collected and Sources" worksheet on the following pages to

1. collect data about the **background conditions** that influence the target population of the community needs assessment.

2. collect data about the **community characteristics** that influence the target population of the community needs assessment.

3. collect data about the **environmental characteristics** that influence the target population of the community needs assessment.

4. collect data about the **socioeconomic characteristics** that influence the target population of the community needs assessment.

5. collect existing data about the **target population** or suggest ideas for collecting new data.

18

☆ Worksheet for Step 3: Data Collected and Sources ☆

Before you start collecting data

1. review the purpose, goals, and objectives of the needs assessment (Step 2).

2. review the issues, characteristics, and questions you have about the target population as listed in the "Background Conditions," "Community Conditions," and "Target Population Data" tables (Step 2).

3. review the methods you have chosen to answer the questions you have about the target population (part of Step 3).

A. Objective or Subjective Data on Background Conditions

- Look back to the section of Step 2 on "**Background Conditions**." Pick at least 2 issues and collect data about those issues. Cite your sources, whether objective or subjective.

☆ Background Conditions ☆
Issue 1: _____
Important data:
Citation:
Issue 2: _____
Important data:
Citation:

B. Objective or Subjective Data about the Community – Community Characteristics

- Look back to the section of Step 2 on "**Community Characteristics**." Pick at least two community characteristics and collect data about those issues. Cite your sources, whether objective or subjective.

☆ Community Characteristics ☆
Community characteristic 1: _____
Important data:
Citation:
Community characteristic 2: _____
Important data:
Citation:

Objective or Subjective Data about the Community – Environmental Characteristics

- Look back to the section of Step 2 on "**Environmental Characteristics**." Pick at least two environmental characteristics and collect data about those issues. Cite your sources, whether objective or subjective.

☆ Environmental Characteristics ☆
Environmental characteristic 1: _____
Important data:
Citation:
Environmental characteristic 2: _____
Important data:
Citation:

Objective or Subjective Data about the Community – Socioeconomic Characteristics

- Look back to the section of Step 2 on "**Socioeconomic Characteristics**." Pick at least two socioeconomic characteristics and collect data about those issues. Cite your sources, whether objective or subjective.

☆ **Socioeconomic Characteristics** ☆
Socioeconomic characteristic 1: _____
Important data:
Citation:
Socioeconomic characteristic 2: _____
Important data:
Citation:

22

C. Objective or Subjective Data about the Community – Target Population

- Look back to the section of Step 2 on **"Target Population Data."** Pick at least one question that you can answer using existing objective data or that you can ask a key informant. Cite your sources, whether objective or subjective.

☆ **Target Population Data** ☆

Question about the target population's nutritional status (from the table at the end of Step 2):

Important data that helps answer the question:

Citation:

Question about the target population's nutritional status (from the table at the end of Step 2):

Important data:

Citation:

☆ Step 4: Analyze and Interpret the Data ☆

Overview

In this section you will analyze and interpret the data you collected and prepare an executive summary based on the guidelines presented on page 59 in your textbook, including 3 or 4 key points revealed by the assessment. You can find examples of these types of executive summaries on pages 103-104 in your textbook.

Activity

Use the "Community Needs Assessment Analysis" worksheet on the next page to

1. interpret the state of the target population's health status, including the severity, extent, and frequency of health problems.

2. describe the health care services and programs that are currently available to assist the target population, including strengths and weaknesses.

3. describe the interaction between the target population's health status and the health status of the surrounding community.

4. summarize how the current target population's nutritional problems relate to the environmental and social characteristics identified during the data collection phase of the assessment.

5. write an executive summary with 3 or 4 key findings from the community needs assessment.

✩ Worksheet for Step 4: Community Needs Assessment Analysis ✩

Look back at all of the data you collected during Step 3 of the community needs assessment and answer the following.

1. What is the state of the target population's health status? Include issues such as the severity, extent, or frequency of health problems.

2. What health care services and programs are currently available to assist the target population? What are the strengths and weaknesses of these resources?

3. How is the health status of the target population related to the health care status of the community or health care available in the larger community (community characteristics)?

4. Summarize how the current target population's nutritional problems relate to the environmental and social characteristics identified during the data collection phase of the assessment.

5. Use your responses to questions 1 to 4 above and other data collected, both objective and subjective, to write an executive summary with 3 or 4 key findings from the community needs assessment.

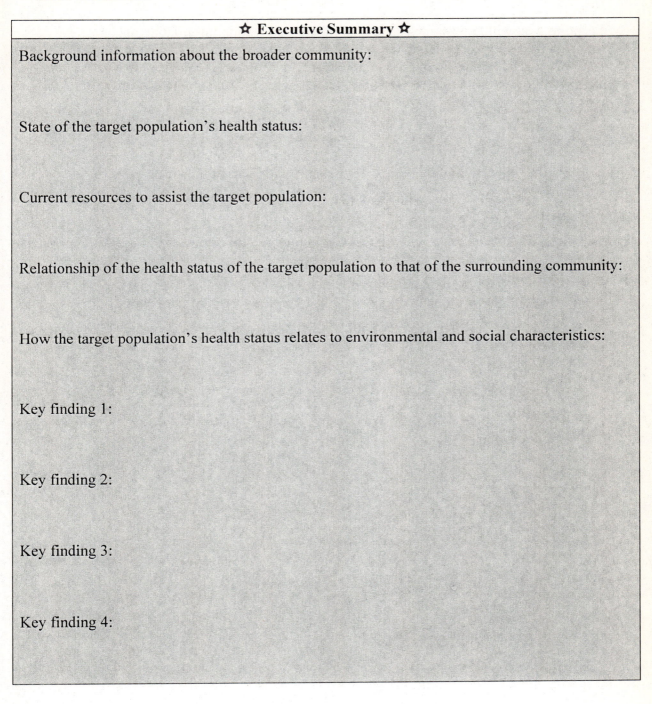

☆ **Executive Summary** ☆

Background information about the broader community:

State of the target population's health status:

Current resources to assist the target population:

Relationship of the health status of the target population to that of the surrounding community:

How the target population's health status relates to environmental and social characteristics:

Key finding 1:

Key finding 2:

Key finding 3:

Key finding 4:

☆ Step 5: Share the Findings of the Assessment ☆

Overview

By this point, you will have done a great deal of work collecting information about the nutritional needs of a target population. It's important to share what you've learned with agencies and organizations in your community. The information is valuable and sharing may help to prevent duplication of efforts. At the same time, you will help increase awareness of important nutrition issues that need attention.

Activity

Use the "Sharing the Community Needs Assessment" worksheet on the next page to

1. list potential key informants who should be contacted for approval before the collected data are shared.

2. list potential stakeholders, agencies, and community organizations who might be interested in the information in the community needs assessment.

3. share the findings of the community needs assessment with at least one stakeholder, agency, or organization. As an alternative, share the assessment with your classmates.

4. write a summary of the feedback provided to you from the stakeholder, agency, organization, or classmates.

☆ Worksheet for Step 5: Sharing the Community Needs Assessment ☆

Let's think about ways that you or your group can share the outcomes of your community needs assessment.

1. First, think back about the key informants who have provided information to you during this assessment process. Are there any key informants you should contact for permission before the needs assessment is shared? If so, list them below and include their contact information. Be sure to contact them for their permission before sharing the outcomes of the assessment.

_____

_____

_____

_____

2. List stakeholders, agencies, and community organizations that might benefit or be interested in your community needs assessment.

_____   _____

_____   _____

_____   _____

_____   _____

_____   _____

3. Set up an appointment to share the community needs assessment with at least one of the stakeholders, agencies, or organizations listed in number 2.

Name: _____

Address: _____

Contact Person: _____

Date and Time of the Appointment: _____

As an alternative, your instructor may ask you to share the outcomes during a brief presentation to your classmates.

28

4. Write a brief summary of the feedback that you obtained while sharing the outcomes of your community needs assessment.

☆ Feedback From Stakeholders, Agency, Organization, or Classmates ☆
Name(s) and address(es) of those who were contacted: Summary of feedback provided:

☆ Step 6: Set Priorities for Next Steps ☆

Overview

During the data collection and sharing processes of the community needs assessment, many important nutrition needs will have been identified. In almost all cases, it will be impossible to address all of these needs at once. The community nutritionist will need to think carefully and realistically to prioritize which nutrition issue needs to be given the highest priority.

In some cases, as a follow-up to the community needs assessment, your instructor may ask you to design, implement (and perhaps market and evaluate) a nutrition education project based on the needs assessment. This section will help you focus on a nutrition education project that can realistically be completed in the time available during the course session or semester.

Activity

Use the "Setting Priorities" worksheet on the next page to

1. list all of the nutrition education needs identified in the community needs assessment.

2. identify the nutrition problem that is the highest priority and needs immediate attention.

3. identify the nutrition problem that you can most realistically address at this time, given time, budget, and resource constraints.

☆ Worksheet for Step 6: Setting Priorities ☆

1. Look back at the data you collected during your community needs assessment. Also consider the feedback you received when you shared the results of your assessment. List all of the nutrition problems and education needs that you uncovered.

_____

_____

_____

_____

_____

_____

_____

_____

_____

_____

_____

_____

_____

_____

_____

_____

_____

_____

_____

_____

2. Look back over the list you made for number 1 on the previous page. What nutrition problem do you feel deserves the highest priority at this time? Justify your response. In addition, look at the *Healthy People 2020* Objectives (Appendix A) and state which one most closely relates to this nutrition problem.

Priority 1 Nutrition Problem:

_____

_____

Justification:

_____

_____

Related *Healthy People 2020* Objective:

_____

3. Again, look back over the list you made for number 1 on the previous page. Think about issues such as time available during the remainder of the course, potential costs and other issues (transportation, schedules), and available resources. Realistically, which nutrition problem could you address at this time? (Your instructor may ask you to design a nutrition education lesson or conduct a more detailed nutrition assessment later in this course based on this response.)

Nutrition Problem:

_____

_____

Why Chosen:

_____

_____

Related *Healthy People 2020* Objective:

_____

☆ Step 7: Choose a Plan of Action ☆

Overview

Congratulations on completing your community needs assessment project! All that remains at this point is to write your final report.

You have invested a great deal of time and effort to define an important nutritional problem, collect important data about that problem, write an executive summary, share the outcomes of your assessment, obtain valuable feedback, and prioritize the nutrition problem that should be given attention.

At this point, a community nutritionist would need to sit down and make an important decision about what to do next. Some potential next steps are to

- use key findings as a jumping off point to affect change in public policy.

- organize a seminar, conference, workshop, or community event to educate others about the nutrition problem and garner further input about the nutrition problem from the community or target population.

- look at existing community programs that are currently addressing the nutrition problem and adapt them to more effectively meet the nutrition need.

- design new educational materials (brochure, handout, website) or an educational program (lesson plan, information table, health fair) to address the nutritional problem.

- write a grant proposal to secure funding to support efforts to educate the target population regarding the nutrition problem.

Activity

Use the "Choose a Plan of Action" worksheet on the next page to

1. brainstorm several specific plans of action that you could implement to improve the nutritional health of the members of your target population based on the results of the community needs assessment.

2. describe your plan of action, listing 2-3 goals for what you hope to do or accomplish (process goals) and 2-3 goals for how you would like your target population's health to improve as a result of the plan's implementation (outcome goals).

33

✩ Worksheet Step 7: Choosing a Plan of Action ✩

1. Brainstorm several specific plans of action that you could implement to improve the nutritional health of the members of your target population based on the results of the community needs assessment.

_____

_____

_____

_____

_____

_____

_____

_____

_____

_____

_____

_____

Choose 1 plan of action and write it here:

_____

_____

_____

_____

34

2. List 2-3 goals for what you hope to do or accomplish (process goals) and 2-3 goals for how you would like your target population's health to improve as a result of carrying out the plan (outcome goals).

What I would like to do to carry out my plan of action to address the identified community nutrition education need:

1.

2.

3.

These are the ways that I hope the target population's health will improve as a result of my efforts to carry out my plan of action:

1.

2.

3.

☆ Writing the Final Report ☆

If requested by your instructor, prepare a final report to hand in with the following headings, which correspond with the gray boxes in this workbook:

Title and/or Brief Description of the Lead Organization

Statement of the Nutritional Problem (Be sure to reference and include the citations at the end of the final report.)

Definition of the Community

Purpose of Assessment

Target Population

Goals and Objectives of the Needs Assessment

Data Collected (Be sure to reference and include the citations at the end of the final report.)

- Background Conditions
- Community Characteristics
- Environmental Characteristics
- Socioeconomic Characteristics
- Target Population Data

Executive Summary

Feedback

References

Sample Assessment Report Rubric

Objective/Criteria	Performance Indicators		
	Needs Improvement	Adequate	Exceptional
Define the Nutritional Problem	(0 points) Statement of the nutritional problem is not included.	(2.5 points) Presents a statement of the nutritional problem, but statement is not concise, motivational, or clear in explaining who is affected and how. Supporting data are not presented or not referenced.	(5 points) Presents a concise, motivational statement of the nutritional problem, including who is affected, how many are affected, its impact on health, etc. Supporting data are presented and referenced.
Set the Parameters of the Assessment	(0 points) Many of the parameters described on pages 42 to 44 in the textbook are missing.	(5 points) A few of the parameters described on pages 42 to 44 in the textbook are included, but some are missing or are not clear or specific.	(10 points) Most of the parameters described on pages 42 to 44 in the textbook are included, and objectives are clear, are measurable, and use a strong verb.
Collect Data	(0 points) Either subjective or objective data are missing.	(5 points) One or two examples of subjective data and three to five examples of objective data are presented.	(10 points) More than two examples of subjective data and more than five examples of objective data are presented.
Analyze and Interpret the Data	(0 points) Executive summary is missing or does not connect to the data collected.	(5 points) A summary is provided that clearly states 3 or fewer key points and these points vaguely relate the data collected.	(10 points) A summary is provided that clearly states at least 4 key points and these points can be supported by the data collected.
Share the Findings of the Assessment	(0 points) Summary not shared.	(2.5 points) Summary was shared with at least one person and that person's feedback is presented in the report.	(5 points) Summary was shared with at least two people and their feedback is presented in the report.
References	(0 points) 5 or fewer references. Not cited within the body of the report; i.e., just a bibliography at the end.	(2.5 points) 5 to 10 references that are cited correctly in the body of the report.	(5 points) Ten or more references that are cited correctly in the body of the report.
Writing, Grammar, Spelling, and Organization	(0 points) Disorganized. Many grammatical or spelling errors. Section headings not used.	(2.5 points) Thoughts aren't as clear as they could be. Somewhat disorganized. Some grammatical or spelling errors. Some section headings are missing.	(5 points) Clearly and concisely written, without grammatical or spelling errors. Section headings help to organize the assessment.
			out of 50

☆ Appendices ☆

Appendix A: Related *Healthy People 2020* Objectives

Appendix B: Data Resources for Assessing Community Needs

Appendix C: Writing Goals and Objectives

Appendix D: Three Sample Community Needs Assessments

1. City-Wide Community Needs Assessment
2. County-Wide Community Needs Assessment
3. State-Wide Community Needs Assessment

☆ Appendix A: Related *Healthy People 2020* Objectives[2] ☆

Food Safety

- FS-1 Reduce infections caused by key pathogens transmitted commonly through food

- FS-2 Reduce the number of outbreak-associated infections due to Shiga toxin-producing *E. coli* O157, or *Campylobacter*, *Listeria*, or *Salmonella* species associated with food commodity groups

- FS-3 Prevent an increase in the proportion of nontyphoidal *Salmonella* and *Campylobacter jejuni* isolates from humans that are resistant to antimicrobial drugs

- FS-4 Reduce severe allergic reactions to food among adults with a food allergy diagnosis

- FS-5 Increase the proportion of consumers who follow key food safety practices

- FS-6 (Developmental) Improve food safety practices associated with foodborne illness in foodservice and retail establishments

Maternal, Infant, and Child Health

- MICH-1 Reduce the rate of fetal and infant deaths

- MICH-2 Reduce the 1-year mortality rate for infants with Down syndrome

- MICH-3 Reduce the rate of child deaths

- MICH-4 Reduce the rate of adolescent and young adult deaths

- MICH-5 Reduce the rate of maternal mortality

- MICH-6 Reduce maternal illness and complications due to pregnancy (complications during hospitalized labor and delivery)

- MICH-7 Reduce cesarean births among low-risk (full-term, singleton, vertex presentation) women

- MICH-8 Reduce low birth weight (LBW) and very low birth weight (VLBW)

- MICH-9 Reduce preterm births

- MICH-10 Increase the proportion of pregnant women who receive early and adequate prenatal care

- MICH-11 Increase abstinence from alcohol, cigarettes, and illicit drugs among pregnant women

- MICH-12 (Developmental) Increase the proportion of pregnant women who attend a series of prepared childbirth classes

- MICH-13 (Developmental) Increase the proportion of mothers who achieve a recommended weight gain during their pregnancies

---

[2] U.S. Department of Health and Human Services. Office of Disease Prevention and Health Promotion. *Healthy People 2020.* Washington, DC. *2020 Topics and Objectives.* Available at http://www.healthypeople.gov/2020/topicsobjectives2020/default.aspx. Accessed March 29, 2012.

- MICH-14 Increase the proportion of women of childbearing potential with intake of at least 400 µg of folic acid from fortified foods or dietary supplements

- MICH-15 Reduce the proportion of women of childbearing potential who have low red blood cell folate concentrations

- MICH-16 Increase the proportion of women delivering a live birth who received preconception care services and practiced key recommended preconception health behaviors

- MICH-17 Reduce the proportion of persons aged 18 to 44 years who have impaired fecundity (i.e., a physical barrier preventing pregnancy or carrying a pregnancy to term)

- MICH-18 (Developmental) Reduce postpartum relapse of smoking among women who quit smoking during pregnancy

- MICH-19 (Developmental) Increase the proportion of women giving birth who attend a postpartum care visit with a health worker

- MICH-20 Increase the proportion of infants who are put to sleep on their backs

- MICH-21 Increase the proportion of infants who are breastfed

- MICH-22 Increase the proportion of employers that have worksite lactation support programs

- MICH-23 Reduce the proportion of breastfed newborns who receive formula supplementation within the first 2 days of life

- MICH-24 Increase the proportion of live births that occur in facilities that provide recommended care for lactating mothers and their babies

- MICH-25 Reduce the occurrence of fetal alcohol syndrome (FAS)

- MICH-26 Reduce the proportion of children diagnosed with a disorder through newborn blood spot screening who experience developmental delay requiring special education services

- MICH-27 Reduce the proportion of children with cerebral palsy born as low-birth weight infants (less than 2,500 grams)

- MICH-28 Reduce occurrence of neural tube defects

- MICH-29 Increase the proportion of young children with an autism spectrum disorder (ASD) and other developmental delays who are screened, evaluated, and enrolled in early intervention services in a timely manner

- MICH-30 Increase the proportion of children, including those with special health care needs, who have access to a medical home

- MICH-31 Increase the proportion of children with special health care needs who receive their care in family-centered, comprehensive, coordinated systems

- MICH-32 Increase appropriate newborn blood-spot screening and follow-up testing

- MICH-33 Increase the proportion of very-low birth weight (VLBW) infants born at level III hospitals or subspecialty perinatal centers

40

Nutrition and Weight Status

- NWS-1 Increase the number of states with nutrition standards for foods and beverages provided to preschool-aged children in child care

- NWS-2 Increase the proportion of schools that offer nutritious foods and beverages outside of school meals

- NWS-3 Increase the number of states that have state-level policies that incentivize food retail outlets to provide foods that are encouraged by the *Dietary Guidelines*

- NWS-4 (Developmental) Increase the proportion of Americans who have access to a food retail outlet that sells a variety of foods that are encouraged by the *Dietary Guidelines for Americans*

- NWS-5 Increase the proportion of primary care physicians who regularly measure the body mass index of their patients

- NWS-6 Increase the proportion of physician office visits that include counseling or education related to nutrition or weight

- NWS-7 (Developmental) Increase the proportion of worksites that offer nutrition or weight management classes

- NWS-8 Increase the proportion of adults who are at a healthy weight

- NWS-9 Reduce the proportion of adults who are obese

- NWS-10 Reduce the proportion of children and adolescents who are considered obese

- NWS-11 (Developmental) Prevent inappropriate weight gain in youth and adults

- NWS-12 Eliminate very low food security among children

- NWS-13 Reduce household food insecurity and in doing so reduce hunger

- NWS-14 Increase the contribution of fruits to the diets of the population aged 2 years and older

- NWS-15 Increase the variety and contribution of vegetables to the diets of the population aged 2 years and older

- NWS-16 Increase the contribution of whole grains to the diets of the population aged 2 years and older

- NWS-17 Reduce consumption of calories from solid fats and added sugars in the population aged 2 years and older

- NWS-18 Reduce consumption of saturated fat in the population aged 2 years and older

- NWS-19 Reduce consumption of sodium in the population aged 2 years and older

- NWS-20 Increase consumption of calcium in the population aged 2 years and older

- NWS-21 Reduce iron deficiency among young children and females of childbearing age

- NWS-22 Reduce iron deficiency among pregnant females

Physical Activity

- PA-1 Reduce the proportion of adults who engage in no leisure-time physical activity

- PA-2 Increase the proportion of adults who meet current federal physical activity guidelines for aerobic physical activity and for muscle-strengthening activity

- PA-3 Increase the proportion of adolescents who meet current federal physical activity guidelines for aerobic physical activity and for muscle-strengthening activity

- PA-4 Increase the proportion of the nation's public and private schools that require daily physical education for all students

- PA-5 Increase the proportion of adolescents who participate in daily school physical education

- PA-6 Increase regularly scheduled elementary school recess in the United States

- PA-7 Increase the proportion of school districts that require or recommend elementary school recess for an appropriate period of time

- PA-8 Increase the proportion of children and adolescents who do not exceed recommended limits for screen time

- PA-9 Increase the number of states with licensing regulations for physical activity provided in child care

- PA-10 Increase the proportion of the nation's public and private schools that provide access to their physical activity spaces and facilities for all persons outside of normal school hours (that is, before and after the school day, on weekends, and during summer and other vacations)

- PA-11 Increase the proportion of physician office visits that include counseling or education related to physical activity

- PA-12 (Developmental) Increase the proportion of employed adults who have access to and participate in employer-based exercise facilities and exercise programs

- PA-13 (Developmental) Increase the proportion of trips made by walking

- PA-14 (Developmental) Increase the proportion of trips made by bicycling

- PA-15 (Developmental) Increase legislative policies for the built environment that enhance access to and availability of physical activity opportunities

42

☆ Appendix B: Online Data Resources for Assessing Community Needs ☆

Name	URL Address
Behavioral Risk Factors Surveillance System (BRFSS)	http://www.cdc.gov/brfss/index.htm
Centers for Disease Control and Prevention (CDC)	http://www.cdc.gov/
CDC – Chronic Disease GIS Exchange	http://www.cdc.gov/dhdsp/maps/gisx/index.html
CDC – School Health Profiles	http://www.cdc.gov/healthyyouth/profiles/index.htm
CDC – Social Determinants of Health Maps	http://www.cdc.gov/dhdsp/maps/social_determinants_maps.htm
CDC – Obesity Trends	http://www.cdc.gov/nccdphp/dnpa/obesity/trend/maps
County Health Rankings	http://www.countyhealthrankings.org/
Community Health Assessment Clearinghouse (New York)	http://www.health.state.ny.us/statistics/chac
Community Health Status Indicators Report	http://www.communityhealth.hhs.gov/HomePage.aspx
Community Nutrition Mapping Project	http://www.ars.usda.gov/Services/docs.htm?docid=15656
DataFerret System	http://dataferrett.census.gov/
Division of Heart Disease and Stroke Prevention – Heart Disease & Stroke Maps	http://apps.nccd.cdc.gov/giscvh2
Economic Research Service – Data Sets	http://www.ers.usda.gov/Data/FoodSecurity/
Economic Research Service – Household Food Security in the U.S. in 2010	http://www.ers.usda.gov/Publications/ERR125/
FedStats	http://www.fedstats.gov/
Feeding America – Map the Meal Gap 2011: Child Food Insecurity	http://feedingamerica.org/hunger-in-america/hunger-studies/map-the-meal-gap.aspx
Food Research and Action Center (FRAC)	http://www.frac.org/
Health Indicators Warehouse	http://healthindicators.gov/
Kaiser Family Foundation State Health Facts	http://www.statehealthfacts.org/
Kids Count	http://kidscount.org/
MEDLINE (National Library of Medicine)	http://www.nlm.nih.gov
Morbidity and Mortality Weekly Report (MMWR)	http://www.cdc.gov/mmwr/
National Agricultural Statistics Service	http://www.nass.usda.gov/
NCHS National Vital Statistics System	http://www.cdc.gov/nchs/nvss.htm
NCHS Press Room	http://www.cdc.gov/nchs/pressroom/Default.htm
National Health Information Center	http://www.health.gov/nhic/

Name	URL Address
National Institute for Occupational Safety and Health (NIOSH)	http://www.cdc.gov/niosh/
The Community Toolbox	http://ctb.ku.edu
University of Michigan's Institute for Social Research Survey Research Center	http://www.src.isr.umich.edu
University of North Carolina's Odum Institute for Social Science	http://www.irss.unc.edu/odum/home2.jsp
U.S. Bureau of Labor Statistics	http://www.bls.gov/
U.S. Census Bureau	http://www.census.gov/
USDA Food Desert Locator	http://www.ers.usda.gov/data/fooddesert/index.htm
WONDER Database	http://wonder.cdc.gov
Your Food Environment Atlas	http://ers.usda.gov/foodatlas/
Youth Risk Behavior Surveillance System (YRBSS)	http://www.cdc.gov/healthyyouth/yrbs/index.htm

States and local governments may also have their own data resources. An index of state and local government websites is available at

http:// www.loc.gov/rr/news/stategov/stategov.html

For example, Massachusetts has the following resources:

- http://www.mass.gov/hed/economic/eohed/dhcd/ and choose "Community Profiles"

- http://www.mass.gov/eohhs/researcher/community-health/masschip/ and click on "Instant Topics"

- http://www.mass.gov/eohhs/gov/departments/dph/ and click on "Registry of Vital Records"

Here's an example from the state of Washington:

- http://depts.washington.edu/commnutr/home/

☆ Appendix C: Writing Goals and Objectives ☆

Students typically find that writing goals and objectives can be challenging, particularly when writing them for their community needs assessment. You may be familiar with writing behavior change goals and objectives for educational lessons or for a client's behavior change. Let's start there and then adapt for this community needs assessment project.

Writing Goals

When you write goals for an educational lesson, you should be thinking broadly about what the learner will accomplish during the lesson. For example, you might be educating a group of people who are newly diagnosed with hyperlipidemia. A goal would be to *increase the participants' knowledge and skills related to healthy eating and physical activity*.

Similarly, when writing goals for the community needs assessment you should think about what you would like to accomplish by going through this process. What do you want to determine or identify in your target population? It's okay if your **goal** statement is broad, general, abstract, and difficult to measure. You'll take care of those issues when you write your **objectives**.

Page 43 of the *Community Nutrition in Action* textbook and Appendix D of this workbook present excellent examples of community needs assessment goals. Here's one that Nicole wrote for her community needs assessment of Sacramento County, CA:

The nutritional status of employees and barriers to achieving improved nutritional status will be identified.

Writing Objectives

In contrast to goals, objectives are narrow, specific, concrete, and measurable. When writing goals for educational lessons, one helpful guide is to make the objectives S.M.A.R.T.: **s**pecific, **m**easurable, **a**ttainable, **r**elevant, and in a realistic **t**ime frame. For your hyperlipidemia clients, here is an example of a S.M.A.R.T objective:

At the end of the one-hour lesson (realistic time frame) on low-fat foods for hyperlipidemia patients (specific), 75% of the participants (measurable) will be able to list 3 examples (attainable) of low-fat alternatives to high-fat foods (relevant to the lesson).

Notice, too, that objectives use active verbs such as *define, choose, compare, arrange, debate*, etc. In this case the active verb is "list."

It's more challenging to write objectives for a community nutrition assessment than for an educational lesson, but let's give it a try.

In Amanda's community needs assessment (in Appendix D) she wrote the following goal and accompanying objectives:

Goal 1: Compare the extent of food insecurity among families in Texas to the extent of food insecurity experienced throughout the rest of the nation.

- *Objective 1: Assess the severity of poverty and food insecurity in Bexar County.*

- ***Objective 2:*** *Compare rates of food insecurity in Bexar County to rates seen in Texas and nationally.*

These are excellent objectives in that they use action verbs such as assess and compare. They are specific to Bexar County, they can be measured (severity and rates can have numbers attached to them), and they are attainable (this data should be available online) and relevant (the objectives help to reach the stated goal). The only thing that Amanda is missing is the "realistic time frame," so let's add that:

- ***Objective 1:*** *During the next 3 weeks, assess the severity of poverty and food insecurity in Bexar County.*

- ***Objective 2:*** *During the next 3 weeks, compare rates of food insecurity in Bexar County to rates seen in Texas and nationally.*

There are more examples of pitfalls to avoid when writing objectives and examples of well-written instructional objectives on page 597 in the *Community Nutrition in Action* textbook. By concentrating on what you hope to accomplish during the assessment (as compared to concentrating on the learner as you would when designing a lesson plan), you should be able to write high-quality S.M.A.R.T outcome objectives for your community needs assessment.

☆ Appendix D: Three Sample Community Needs Assessments ☆

During their community nutrition course, students enrolled in the online Master of Public Health in Nutrition Program at the University of Massachusetts Amherst and UMassOnline completed the following community needs assessments as part of their semester-long nutrition education project. The students used the information gathered during these assessments to design, market, implement, and evaluate community nutrition education interventions in their target communities.

1. City-Wide Community Needs Assessment

Title: Community Needs Assessment of Working Adults in Sacramento, CA, October 2009

Prepared by Nicole Geurin, RD
Online MPH in Nutrition Student, University of Massachusetts Amherst

Statement of the Nutritional Problem:

In Sacramento, like in the rest of the United States, the prevalence of overweight and obesity in working adults is a growing concern (Behavioral Risk Factor Surveillance System City and County Data, 2008). A large percentage of the health care costs associated with overweight and obesity falls on the shoulders of the employer (Chenoweth, 2005). Therefore, the implementation of wellness programs in the workplace is a growing trend in the U.S. However, little data is known about the types of worksite wellness programs implemented in Sacramento worksites, or the effectiveness of these programs in improving healthful behaviors and reducing the prevalence of overweight in Sacramento adults.

Definition of the Community: The city of Sacramento

Purpose of Assessment:

1. To obtain information about the nutritional status of Sacramento employees and determine if a worksite wellness program would be of value.

2. To identify the types of wellness programs implemented in Sacramento worksites, and to evaluate the effectiveness of these programs in improving healthful behaviors and reducing the prevalence of overweight in Sacramento adults.

Target Population: Working adults in Sacramento

Overall Goal of Assessment: The nutritional status of employees and barriers to achieving improved nutritional status will be identified

Objectives of Assessment: On a sample of adult employees, within the next few weeks:
- Assess the health status of employees
- Assess the current health-related behaviors of employees
- Identify the perceived barriers to achieving better health of employees
- Determine an effective and feasible nutrition education project to address these issues

Data Collected

Background Conditions

Health behaviors of adults in Sacramento:
- 54.2% of Sacramento adults reported eating fewer than 5 servings of fruits and vegetables per day (California Health Interview Survey, 2005)
- Self-reported physical activity levels: no physical activity: 14.9%, some physical activity: 47.9%, moderate physical activity: 19%, vigorous physical activity: 18.1% (California Health Interview Survey, 2007)
- Self-reported number of times fast food was eaten in past week: No times: 33.4%, one time: 32.1%, two times: 16%, three times: 9%, four or more times: 9.6% (California Health Interview Survey, 2007)
- 15.4% reported engaging in binge drinking in the past month (California Health Interview Survey, 2005)

Community Characteristics

Sacramento Census Data:
- Sacramento County 2008 population estimate: 1,394,154 (FedStats, 2008)
- Percentage of adults between the ages of 18 and 65 years: 62.7% (FedStats, 2007)
- Age-adjusted death rate (age-adjusted # of deaths per 100,000) due to cancer, Sacramento County: 176.1 (County Health Status Profiles, 2008)
  - *Healthy People 2010* Objective: 158.6 (County Health Status Profiles, 2008)
- Number of private non-farm establishments: 28,823 (FedStats, 2006)
- Number of firms: 90,876 (FedStats, 2002)

Characteristics of working adults in Sacramento:
- The number of employed residents in the city of Sacramento was 199,085 in 2008. If you include the employed residents in Sacramento suburbs, that number jumps to 983,616. (State of the Cities Data Systems Current Labor Force Data, 2008)
- Mean travel time to work for Sacramento residents aged 16+ is 23.4 minutes. (FedStats, 2000)

Environmental Characteristics

Wellness initiatives at Sacramento City Unified School District include: a workout facility, walking clubs, health fairs, and lunch and learn seminars. Most of these initiatives take place at the main building (Serna Center). Lack of education is a large barrier to eating healthfully for employees. A series of nutrition presentations in previous years was very well received by employees. One of the biggest problems in reaching all employees is that they are so spread out at multiple sites across the city. (Clemmens, Marianne, Director of Employee Risk Services at Sacramento City Unified School District, personal conversation, October 7th, 2009)

Socioeconomic Characteristics
- Income per capita: $35,197 (FedStats 2006)
- Demographics: Percent white not Hispanic: 52.6%, percent Hispanic or Latino: 19.6%, percent Asian: 13.3%, percent Black: 10.5% (FedStats, 2007)
- Education Attainment – Sacramento County (SNAPS Data, 2003):
  - No School: 18,042
  - No High School: 33,632

48

- o   Some High School: 77,596
- o   High School: 176,525
- o   Some College: 205,947
- o   Associate's Degree: 69,105
- o   College Degree: 129,263
- o   Master's Degree: 40,171
- o   Professional Degree: 16,793
- o   Doctoral Degree: 5,414

Target Population Data

A survey of 25 employees at Holt of California (Geurin, 2008) indicated the following barriers to eating healthfully, in perceived order of importance:

1. I don't have time to purchase and prepare healthy foods
2. Healthy foods are too expensive
3. I don't always know which foods are healthy and which are not
4. It's too difficult to change my eating habits to ones that are healthier
5. Healthy foods are too difficult to find
6. I don't like the taste of most healthy foods

Subjective Data from RDs Involved in Corporate Wellness:

The following questions were asked of RDs on the Nutrition Entrepreneurs Corporate Wellness listserv, with the following responses received:

1. In your opinion, what are some of the greatest nutritional problems and/or barriers to eating healthfully for working adults today?

   - "Time, or perception of time (lack thereof). Many working adults work long hours, and many have children, eldercare or other responsibilities that make healthy eating and preparing, shopping etc. seem difficult. On the flipside, Americans watch 3-4 hours of TV a day, and some of this time could be better utilized for exercise, better rest, and preparing healthy meals." (Jacobs, 2009)

   - "Schedules, the majority of my clients complain about being on the go all the time, having too much work, family routines and functions and effecting [sic] their eating patterns and habits." (Romero, 2009)

   - "I agree with both previous responses, but I still think both Portion and Calorie distortion are major contributors. I developed and taught a weight management class and people were shocked at actual calorie contents of foods and the trend of the increases in potion size, in addition the lack of REAL food in our 'fast-paced/convince' society (everything is made in a factory and opened from a package)- what happened to real food." (O'Connor, 2009)

2. What type of one-time intervention do you think would be most effective in addressing these problems?

   - "Most effective is tough, because people learn differently and are ready for change at different points in time, but this could be addressed effectively in a webinar, or workshop, offering employees quick ways of deciding what to purchase and how often, quick to prepare meals, easily portable snacks etc, and demonstrating how it takes as much time to drive to the takeout place and wait for your takeout, as it does to make a stir fry at home." (Jacobs, 2009)

- "One-on-one counseling discussing meal planning." (Romero, 2009)

- "I think you could do a lecture that addresses a lot of these issues, healthy, quick food options and preparation, planning ahead and portion control." (O'Connor, 2009)

Executive Summary

Over half of Sacramento's residents are overweight or obese. (Behavioral Risk Factor Surveillance System City and County Data, 2008). Unhealthy nutritional habits are likely contributing to overweight in this population (California Health Interview Survey, 2005-2007). The vast majority of working-age adults in Sacramento are employed (State of the Cities Data Systems Current Labor Force Data, 2008). Therefore, worksites are a viable setting for nutrition intervention. Some worksites in Sacramento have some type of wellness program in place; however, the span of these programs may be limited in large organizations (Clemmens, 2009). One of the greatest perceived barriers among Sacramento employees to achieving a healthy diet is lack of time to purchase and prepare healthy meals (Geurin, 2008; Jacobs 2009; Romero 2009). Expense of healthy food and lack of nutrition knowledge may also be key barriers (Geurin 2008; Clemmens 2009; O'Connor 2009). An intervention in the worksite that addresses these barriers may be an effective means of improving healthy nutritional habits and, ultimately, decreasing the prevalence of overweight in this population.

Feedback

The findings were shared with two employees that manage worksite wellness campaigns, one from the Health Education Council and one from Kaiser Permanente. They agreed with the findings and added that stress management was another topic of concern for employees, and showing how nutrition and physical activity can help lower stress levels is a good way to approach the topic. Excessive "eating out" and over-sized portions were identified as other nutritional problems. They also suggested a visual or interactive presentation intervention, such as a cooking demo or cost analysis of a healthy home-cooked meal. (Peterson 2009; Tompkins 2009)

References

1. Behavioral Risk Factor Surveillance System City and County Data (2008). CDC. Accessed on 9/29/09 from: http://apps.nccd.cdc.gov/BRFSS-SMART/MMSARiskChart.asp?yr=2008&MMSA=285&cat=OB&qkey=4409&grp=0
2. California Health Interview Survey (2005-2007). UCLA Center for Health Policy Research. Accessed 10/3/09 from: http://www.chis.ucla.edu/
3. Chenoweth, D. (2005). *The Economic Costs of Physical Inactivity, Obesity and Overweight in California Adults: Health Care, Workers' Compensation and Lost Productivity*. Accessed on 10/6/09 from www.cdph.ca.gov/HealthInfo/healthyliving/nutrition/Documents/CostofObesityToplineReport.pdf
4. Clemmens, Marianne. Director of Employee Risk Services at Sacramento City Unified School District, personal conversation, October 7th, 2009.

5. County Health Status Profiles (2008). California Department of Public Health and California Conference of Local Health Officers. Accessed on 10/7/09 from: http://www.caldiabetes.org/content_display.cfm?contentID=1195&ProfilesID=272&CategoriesID=0

6. FEDStats (2000-2009). Bureau of Labor Statistics, Federal Bureau of Investigation, National Oceanic and Atmospheric Administration, U.S. Census Bureau, U.S. Department of Housing and Urban Development. Last Revised: Friday, 10-Jul-2009 13:30:46 EDT. Accessed on 9/29/09 from: http://fedstats.gov/qf/states/06/0664000.html

7. Geurin, Nicole (2008). Survey: Perceived Barriers to a Healthful Diet by Employees at Holt of California.

8. O'Connor, deNelle (2009). Global Wellness Director of Plus One Health Management. Response on Nutrition Entrepreneurs Corporate Wellness Listserv, 10/8/2009.

9. Jacobs, Elysa (2009). Response on Nutrition Entrepreneurs Corporate Wellness Listserv, 10/8/2009.

10. Romero, Natalie (2009). Wellness Advantage and Community Nutrition RD at Baptist Health Medical Plaza. Response on Nutrition Entrepreneurs Corporate Wellness Listserv, 10/8/2009.

11. Peterson, Eileen (2009). Senior Health Educator, Kaiser Permanente. Email correspondence, October 9, 2009.

12. SNAPS Data (2003). CDC SNAPS, California, Sacramento County. Accessed on 10/7/09 from: http://emergency.cdc.gov/snaps/data/06/06067.htm

13. State of the Cities Data Systems Current Labor Force Data (2008). Output for Sacramento, CA. Accessed on 10/3/09 from: http://socds.huduser.org/BLS_LAUS/OUTPUT.odb

14. Tompkins, Mai Linh (2009). Program Administrator, Health Education Council. Personal Conversation, October 9, 2009.

2. County-Wide Community Needs Assessment

Title: Nutrition Assessment of Bexar County, TX, October 2008
The Problem of Hunger Among Residents of Bexar County

Prepared by Amanda Sylvie, RD
Online MPH in Nutrition Student, University of Massachusetts Amherst

Nutrition Problem

Texas is one of the most food insecure states in the nation. With 21% of the Bexar County population living under 125% of the Federal Poverty Line (Food Research and Action Center [FRAC] Survey, 2006), single women with children constitute the majority of this population and more frequently experience food insecurity (U. S. Census Bureau [Census] S1703), (Census S1702, 2006). Food insecurity and hunger is known to impair cognitive development in children and increase the frequency of obesity, with subsequent health-related consequences, in women (Weinreb, 2002; Townsend, Peerson, Love, Achterberg, Murphy, 2001). The Supplemental Nutrition Assistance Program participation rates in Bexar County are low (FRAC City-by-City, 2005), and families do not have sufficient resources (food, money, or budgeting knowledge) to avoid relying on emergency food assistance.

Location of Interest: Bexar County, Texas of which San Antonio is the largest metropolitan city.

Community

The indigent population living at less than 130% of the Federal Poverty Line (FPL) within the metropolitan area of San Antonio and other municipalities within Bexar County, Texas.

Purpose

Examine the extent of food insecurity in Bexar County and the effectiveness of major food assistance programs in relieving hunger.

Target Population

Bexar County families who are 130% below the FPL.

Goals and Objectives of the Needs Assessment

Goal 1: Compare the extent of food insecurity among families in Texas to the extent of food insecurity experienced throughout the rest of the nation.
- Objective 1: Assess the severity of poverty and food insecurity in Bexar County.
- Objective 2: Compare rates of food insecurity in Bexar County to rates seen in Texas and nationally.

Goal 2: Locate organizations that provide food assistance in Bexar County.
- Objective 1: Identify the roles of various types of local food assistance agencies.
- Objective 2: Determine the trends of SNAP benefit participation in Bexar County.
- Objective 3: Compare SNAP usage in Bexar County to the use of SNAP benefits in the rest of Texas and nationally.

Supporting Data

Background Conditions

Issue 1: The number of persons in Bexar County who are eligible to participate in federal or state food assistance programs.
- Texas is the 4th in the nation for worst poverty rate (Kaiser Profile, 2007; Kaiser Compare, 2007)
  - 22% of Texans are <100% FPL (Federal Poverty Line)
  - 21% are between 100 and 199% FPL
  - (U.S. rates are 17% and 19%, respectively)
- 21% of Bexar county is <125% FPL (Cooper, 2007)
- Estimated 40% of the population at <125% FPL may experience food insecurity (FRAC Survey, 2006)

Issue 2: Local programs and types of services involved in anti-hunger efforts.
- The San Antonio Food Bank (SAFB) partners with 280 community agencies that provide emergency food relief. (SAFB, 2008)

Community Characteristics
Characteristic 1: Demographic Data & Trends

Bexar County residents in poverty:
- Come from broken homes
- Typically have a single female as the sole supporter

- Are of a minority race
- Have no high school diploma or GED
- Have a disability
- Do not have a job that provides reliable income
- Have 3 or more children in the family (Census S1703, 2006; Census S1702, 2006)

Characteristic 2: Existing community services and programs

- The SAFB partners with 280 community agencies
- Supplemental Nutrition Assistance Program (SNAP)
- National School Lunch Program (NSLP)
- Special Supplemental Nutrition Program for Women, Infants, and Children (WIC) (SAFB, 2008)

Environmental Characteristics

Characteristic 1: Food Systems

- There are 19 in-state food banks (Texas Food Bank Network)
- The SAFB has widespread coverage of most zip codes within the San Antonio city limits (M. Werneli, personal communication: SNAP-Ed Coordinator, TX DHHS)

Characteristic 2: Transportation Systems

- VIA Bus System – San Antonio's metropolitan bus transportation system

Socioeconomic Characteristics

Characteristic 1: Sociocultural Data & Trends

Those at less than 130% FPL are most frequently:

- Single female as the sole supporter or wage earner
- Minority race
- Did not complete high school
- Have a disability
- Unreliable income
- Have 3 or more children (Census S1703, 2006; Census S1702, 2006)

Characteristic 2: Economic Data & Trends

- 21% of the Bexar County population live below 125% of the FPL (FRAC Survey, 2006)
- 24.2% is the child poverty rate in Bexar County (Cooper, 2007)
- 20% of the Bexar County population qualifies for SNAP benefits (Eschback, 2006; FRAC Survey, 2006)
    - 66% of these are families with children
    - 7% of this population are elderly (Powers, 2007)

Target Population Data

What coping methods are being used by families when their resources run out?

- In addition to receiving assistance from community agencies, other coping methods include turning to extended family, friends, local churches, and other resources (such as pawn shops to sell possessions). (N. Stephens, personal communication, October 1, 2008)

Analysis and Interpretation of the Data

Texas is one of the most food-insecure states in the nation (Valls-Trelles, 2000) with 21.1% of the households being either food insecure or of very-low food security (Cooper, 2007). The primary reason for food insecurity nationwide is poverty (Sodexho, 2007). With 22% of Texans below 100% of the Federal Poverty Level (FPL) and 21% between 100 and 199% of the FPL, compared to the U.S. respective rates of 17% and 19%, Texas ranks fourth in the nation for having the worst poverty rate (Kaiser State Health Facts [Kaiser] Profile, 2007; Kaiser Compare, 2007). A total of 324,850 people (or 21% of the population) of Bexar County exist below 125% of the Federal Poverty Line (FPL), indicating 40% of this population may struggle with food insecurity (Cooper, 2007; FRAC Survey, 2006). The Texas child poverty rate is 24.9%, the 43rd worst in the nation, with 1,548,069 children in poverty (Cooper, 2007). In Bexar County, the child poverty rate is 24.2%, which means almost one in four children in the county are likely food insecure (FRAC Survey, 2006). Hunger not only has wide-ranging effects on children's physical and cognitive development and mental health, but can also impact future health status. Children who experience hunger are more likely to have higher anxiety, "significantly higher chronic illness counts," and "internalizing behavior problems" (Weinreb et al., 2002). Alaimo (2000) notes that children with insufficient food intake may have lower scores in arithmetic and have a higher incidence of grade repetition. Studies have shown parents, typically single mothers with children, are frequently able to protect their children from food insecurity with hunger by going without food themselves (Kaiser, 2002). Linked with the practice of protecting their children from excess hunger, women who experience food insecurity are more likely to exhibit an overweight status (Townsend et al, 2001), leading to costly future health problems such as cardiovascular disease and diabetes.

In addition to federal food assistance programs that include the Supplemental Nutrition Assistance Program (SNAP; previously known as the Food Stamp Program), the National School Lunch Program (NSLP), and the Special Supplemental Nutrition Program for Women, Infants, and Children (WIC), the San Antonio region that includes Bexar county has at least 280 community agencies that partner with the San Antonio Food Bank to bring food relief to the hungry children, families, and elderly (SAFB, 2008). Nancy Stephens, director of The Agape Ministry, one of the community agencies that partners with the San Antonio Food Bank, described how community agencies promote access to federal assistance. Community agencies refer clients to the San Antonio Food Bank by phone while at the agency's location. After collecting the client's information, the food bank agent mails the client an application. Once the application is returned by mail or in person, the food bank processes it and follows up with the client to ensure they receive assistance from the SNAP, WIC, Medicaid, Medicare or other programs for which they are eligible (N. Stephens, personal communication, October 1, 2008).

SNAP participation rates in Bexar County are low. Though there was a 65.6% increase in enrollment between the years 2000 and 2005, only 59% of eligible persons in Bexar County participate in SNAP (FRAC City-by-City, 2005) compared to the U.S. average participation rate of 61.5% (Vollinger, 2004). Within Bexar County, approximately 20% of the population qualifies for SNAP benefits (Eschback, 2006; FRAC Survey, 2006), and of the households receiving benefits, 66% have children and 7% have elderly persons (Powers, 2007).

Bexar County is home to San Antonio, the third largest city in Texas. Urban populations throughout the U.S. are known for having higher rates of poverty than smaller towns and rural locations. In association with greater poverty rates, the population in urban areas pays higher food prices and experiences more hunger and food insecurity (FRAC City-by-City, 2005). The majority of Bexar County residents experiencing poverty come from broken homes with a single female being the sole supporter, are of a minority race, did not complete high school, have a

54

disability, or do not have a job providing reliable income. Those suffering the greatest levels of poverty are children from families with three or more children (Census S1703, 2006; Census S1702, 2006). Similarly, Nancy Stephens reports that the majority of families who seek food assistance from The Agape Ministry have at least one wage-earner who works a minimum wage job and receives SNAP or WIC support. It is not uncommon for such families to have five or six children, making it difficult for the family to retain enough resources to complete a month with adequate food. In addition to receiving assistance from community agencies when resources run out, Stephens noted that other coping methods include turning to extended family, friends, local churches, and other resources such as pawn shops to sell possessions (N. Stephens, personal communication, October 1, 2008).

With the potential for 123,074 Bexar County families to be food insecure (Census S1702, 2006; Nord, Andrews, Carlson, 2006), the importance of federal and community food assistance programs is undeniable. Outreach efforts such as those promoted by Marc Wernli as the SNAP-Ed Coordinator for the Texas Department of Human Services include: contracting with the nineteen in-state food banks, providing assistance with applications for SNAP, and cross-training Children's Health Insurance Program (CHIP) personnel who interact regularly with low-income clients to promote SNAP (M. Wernli, personal communication, September 30, 2008). These outreach efforts and agency partnering have improved the indigent population's access to resources and benefits. To significantly decrease hunger related to poverty, continued effort is needed so more families will participate in food assistance programs (i.e., SNAP, NSLP, and the Summer Food Service Program [SFSP], etc.).

Conclusion

Even with assistance provided through SNAP and other programs, many of those in poverty still do not have the ability to make it through an entire month without seeking emergency food sources. They either have insufficient funds, insufficient food, or insufficient knowledge, to include practical ways of extending the food or money they have, to promote food security while avoiding emergency food assistance. The USDA's Cooperative Extension Service provides the Expanded Food and Nutrition Education Program (EFNEP) in San Antonio (Jacqueline Replogle, personal communication, September 29, 2008), but the education efforts do not reach an adequate number of families to significantly decrease food insecurity throughout Bexar County. Similarly, SNAP-Ed (formerly Food Stamp Nutrition Education) education efforts are limited by the number of families it impacts. To effectively lower rates of food insecurity in Bexar County, targeted behavior change among beneficiaries of food assistance must occur in areas of nutritious and frugal grocery shopping, careful meal planning, healthful food preparation, and monthly budgeting.

Executive Summary

Texas is one of the most food insecure states in the nation. With 21% of the Bexar County population living under 125% of the Federal Poverty Line (Food Research and Action Center [FRAC] Survey, 2006), single women with children constitute the majority of this population and more frequently experience food insecurity (U. S. Census Bureau [Census] S1703; Census S1702, 2006). Food insecurity and hunger are known to impair cognitive development in children and increase the frequency of obesity, with subsequent health-related consequences, in women (Weinreb, 2002; Townsend, Peerson, Love, Achterberg, Murphy, 2001). The SNAP participant rates in Bexar County are low (FRAC City-by-City, 2005), and families do not have sufficient resources (food, money, or budgeting knowledge) to avoid relying on emergency food assistance.

Major findings include:

- As the primary cause of food insecurity, poverty affects 35.4% of Bexar County children and 24.1% of the Bexar County population who live at or below 50-100% of the Federal Poverty Line (FRAC Survey, 2006).

- The San Antonio Food Bank, which assists Bexar County and surrounding areas, partners with 280 community agencies that provide emergency food relief (San Antonio Food Bank [SAFB], 2008).

- Only 59% of eligible persons participate in the Supplemental Nutrition Assistance Program, and there was a 65.6% increase in enrollment between the years 2000 and 2005 (FRAC City-by-City, 2005).

- Local agencies indicate the majority of their clients work minimum wage jobs and receive SNAP or WIC benefits, but are unable to make their monetary and food resources extend throughout an entire month (N. Stephens, personal communication, October 1, 2008).

To significantly decrease hunger related to poverty, communities need to continuously strengthen efforts to educate indigent persons of available food assistance, and enroll more families in food assistance programs such as SNAP, the National School Lunch Program, and the Summer Food Service Program. By promoting and providing education in areas of nutrition, meal planning and preparation, and budgeting to low-income women, Bexar County could realize a lower frequency of food insecurity with less dependence upon emergency food relief in addition to improved cognition and health among its women and children.

Feedback

After discussing my findings, I asked Ms. Stephens what "one thing" was needed most by those who utilize the services of The Agape Ministry. She responded that the families and individuals needed to learn simple budgeting methods so they can make the best use of the monetary resources available to them. (N. Stephens of The Agape Ministry)

References

1. Valls-Trelles, P., Data from US Census Bureau, Detailed Poverty Tables: 2000 (Released September 2001) and US Dept. of Agriculture, Food & Nutrition Research, Food Security in the United States: How Many Households? How Many People?, 1996-1998 (Released August 2000). Retrieved September 25, 2008, from http://www.endhunger.com/assets/States/Texas.html
2. Cooper, R., (May 22, 2007), *Food Research and Action Center: State of the States 2007*; Retrieved September 25, 2008, from http://www.frac.org/State_Of_States/2007/states/TX.pdf
3. Food Research and Action Center (FRAC) performed analysis; Data from U.S. Census Bureau, American Community Survey 2006; Retrieved September 26, 2008 from http://www.frac.org/data/county/ACS2006_Texas.xls
4. FRAC; Food Stamp Access in Urban America: A City-by-City Snapshot; Retrieved Oct 3, 2008 from http://www.frac.org/pdf/cities2005.pdf

56

5. Powers, S., (October 23, 2007), *Food Research and Action Center: Food Stamp Access in Urban America: A City-by-City Snapshot*; Retrieved September 25, 2008 from http://www.frac.org/pdf/UrbanFoodStamp07.pdf

6. Sodexho (December 2007), *A Status Report on Hunger and Homelessness in America's Cities: A 23-City Survey*; Retrieved September 25, 2008 from http://usmayors.org/HHSurvey2007/hhsurvey07.pdf

7. Kaiser, L., (2002). Food Security and Nutritional Outcomes of Preschool-Age Mexican-American Children. *J Am Diet Assoc*. 2002; 102:924-929.

8. Linda Weinreb, MD, Cheryl Wehler, Jennifer Perloff, MPA, Richard Scott, PhD, David Hosmer, PhD, Linda Sagor, MD and Craig Gundersen, PhD, (2002) Hunger: Its Impact on Children's Health and Mental Health, *Pediatrics* Vol. 110 No. 4 October 2002, pp. e41

9. Alaimo, K., Olson, C., Frongillo, E. (2000) Food Insufficiency and American School-Aged Children's Cognitive, Academic, and Psychosocial Development, *Pediatrics* Vol. 108 No. 1 July 2001, pp. 44-53.

10. Marilyn S. Townsend, Janet Peerson, Bradley Love, Cheryl Achterberg and Suzanne P. Murphy, Food Insecurity Is Positively Related to Overweight in Women, *Journal of Nutrition*. 2001; 131:1738-1745

11. San Antonio Food Bank (SAFB); Current Member Agencies; Retrieved September 25, 2008 from http://safoodbank.org/member_agencies.html

12. U.S. Census Bureau, 2006 American Community Survey, Bexar County, TX: *S1703. Selected Characteristics of People at Specified Levels of Poverty in the Past 12 Months*; Retrieved October 2, 2008 from http://www.factfinder.census.gov/servlet/STTable?_bm=y&-geo_id=05000US48029&-qr_name=ACS_2006_EST_G00_S1703&-ds_name=ACS_2006_EST_G00_&-_lang=en&-redoLog=false&-CONTEXT=st

13. U.S. Census Bureau, 2006 American Community Survey, Bexar County, TX: S1702 Poverty Status in the Past 12 Months of Families; Retrieved October 2, 2008 from http://www.factfinder.census.gov/servlet/STTable?_bm=y&-geo_id=05000US48029&-qr_name=ACS_2006_EST_G00_S1702&-ds_name=ACS_2006_EST_G00_&-_lang=en&-redoLog=false&-CONTEXT=st

14. Kaiser State Health Facts, Texas: Distribution of Total Population by Federal Poverty Level, states (2006-2007), U.S. (2007); Retrieved October 2, 2008 from http://www.statehealthfacts.kff.org/profileind.jsp?ind=9&cat=1&rgn=45

15. Kaiser State Health Facts, Distribution of Total Population by Federal Poverty Level, states (2006-2007), U.S. (2007) ; Retrieved October 5, 2008 from http://www.statehealthfacts.kff.org/comparebar.jsp?ind=9&cat=1&sub=2&yr=85&typ=2&o=a&sort=22

16. Vollinger, E. (2004), *Food Stamp Participation Access Rates State-by-State*; Retrieved October 5, 2008 from http://www.frac.org/html/federal_food_programs/FSP/Participation_Rates_03.html

17. Eschbach, K., *Texas State Data Center Population Estimates Program*; Retrieved October 5, 2008 from http://txsdc.utsa.edu/tpepp/2006_txpopest_county.php

18. Nord, M., Andrews, M., and Carlson, S. (October 30, 2006), *Household Food Security in the United States, 2005*; Retrieved October 3, 2008 from http://www.ers.usda.gov/Publications/ERR29/ERR29b.pdf

3. State-Wide Community Needs Assessment

Title: Community Needs Assessment of the State of New York, June 2010
Ounce of Prevention: Non-profit organization for obesity prevention research and initiatives

Prepared by Denine Stracker, RD
Online MPH in Nutrition Student, University of Massachusetts, Amherst

Statement of the Nutritional Problem

Obesity has become a worldwide epidemic and here in America childhood obesity has reached epic proportions. In fact, the prevalence of obesity among 6- to 11-year old American children has ballooned from approximately 7% in the 1970s to more than 19.6% by 2007-2008 (Ogden et al., 2010).

Obesity rates in New York mirror those of the nation with 20% of elementary school children in New York State obese while only 38% of youth in these grade levels are meeting current physical activity recommendation levels (Health Kids, Healthy NYS, 2009). Children of ethnic minorities are disproportionately affected by obesity and overweight. Hispanics and Native American children of both genders along with African American girls share the highest rates of overweight in the nation. Second and third generation children of Pacific Islander and Asian American families are becoming more susceptible (Crawford, 2001).

New York is home to a very diverse population of 60% White persons, 17% Black persons, 16% persons of Hispanic origin and 7% Asian Americans. In fact, 20% of New Yorkers are foreign-born and 28% speak a language other than English in the home (U.S. Census Bureau, State and County Quick Facts, 2009). The minority population is growing even greater as the percentage of Hispanic New Yorkers increased nearly 30% between 1990 and 2000 whereas White non-Hispanics were the only group to decrease in proportion by -5.6% during the same time period (New York State Minority Health Surveillance).

The obesity rate for White non-Hispanic elementary school children in New York State is 18%, yet 23% of African Americans and 30% of Hispanics are classified as obese (Upstate NY, Grade 3 Oral Health, Physical Activity, and Nutrition Survey). The percentages of New York high school students considered obese are 15% of Hispanics and 11% of Black non-Hispanics compared with only 9% of White non-Hispanics (New York State Minority Health Surveillance Report, 2007). These rates greatly exceed the U.S. NHANES 1999-2002 survey rate of 16% and the *Healthy People 2010* target of 5% (*Healthy People 2010*). Additionally, only 87% of Black non-Hispanic and 88% of Hispanic high school students reported vigorous or moderate activity during the past week using data from the Youth Risk Behavior Survey of 2005.

Definition of the Community

New York State including the New York City boroughs of Manhattan, Bronx, Queens, Staten Island, and Brooklyn

Purpose of Assessment

To obtain information regarding the obesity and overweight statuses among adolescents of minority populations and assess the need for community initiatives

Target Population

Our target population is children of minority backgrounds between the ages of 5 and 12 years living in New York State.

Goals and Objectives of the Needs Assessment

Overall Goal: To identify the knowledge, behavior, and perspectives of parents and children related to overweight and obesity. Environmental factors as well as existing initiatives designed to reduce prevalence and risk factors associated with the disease will be identified.

Goal A: Ascertain data regarding childhood obesity prevalence and identify contributing factors for obesity
- Objective 1: Identify and compare the physical activity levels and nutritional intake among adolescents in NYS and the U.S.
- Objective 2: Identify and compare the NYS and U.S. rates of obesity and overweight among both adults and children of African American, Hispanic, Asian, and Caucasian backgrounds
- Objective 3: Identify health risks and economic costs associated with childhood and adult obesity

Goal B: Research local community organizations dedicated to preventing childhood obesity
- Objective 1: Identify community coalitions and organizations aimed at childhood obesity prevention among NYS children
- Objective 2: Identify local organizations assisting children from ethnic minority households
- Objective 3: Identify organizations promoting nutrition education and physical activity

Background Conditions

- Whitaker et al. found that approximately 80% of children who were overweight at age 10-15 years were obese adults at age 25 years (Whitaker, 1997).

- Approximately one-half of overweight school-aged children remain overweight as adults (Serdula, 1993).

- Obesity in both children and adults is associated with a number of health risks. As Finkelstein, Trogdon, Cohen & Dietz (2009) state, medical spending for the obese across all payers, per capita, is $1,429 higher per year, or roughly 42% higher, than for someone of normal weight. These results indicate that obesity is associated with a 9.1% increase in annual medical spending, compared with 6.5% in 1998.

- Authors Lauer et al. found that obesity acquired in childhood is predictive of adult obesity and development of coronary artery calcification (Lauer, Clarke, & Burns, 1997).

- Additional health problems such as high rates of hypertension, asthma, musculoskeletal discomfort, shortness of breath, and obstructive sleep apnea have all been reported in overweight children (Waters, & Baur, 2003).

- Obese children also suffer from low self-esteem and significantly higher rates of sadness, loneliness, and nervousness, and are more likely to engage in high-risk behaviors such as smoking or consuming alcohol (Strauss, 2000).

- Therefore the key strategy, recognized by most professionals, for controlling the obesity epidemic is prevention (Muller & Mast, 2001).

- As daSilva-Sanigorski et al., (2010) report, because "once present, obesity is extremely difficult to overcome, children are now considered the priority population for interventions to prevent obesity."

- *Healthy People 2010* named overweight and obesity as one of the leading health indicators and included an objective to reduce overweight and obesity in children and adolescents as part of *Healthy People 2020* (*Healthy People 2010*).

- 300,000 deaths each year in the United States are associated with obesity (Office of the Surgeon General).

Community Characteristics

- New York State is composed of 20 million people with 23% of these New Yorkers under the age of 19 years old. New York is home to a very diverse population of 60% White persons, 17% Black persons, 16% persons of Hispanic origin, and 7% Asian Americans. Twenty percent of New Yorkers are foreign-born and 28% speak a language other than English in the home (U.S. Census Bureau. State and County Quick Facts, 2009).

- According to the 2006-2008 U.S. Bureau of Labor and Statistics Data, 14% of New Yorkers live below 100% of the poverty line while 32% live under 200% of the national poverty line (New York State Department of Health Population Survey).

- Ten percent of New Yorkers are supported by SNAP benefits (USDA Agriculture Research Service).

- According to the U.S. Census Bureau 2005 American Community Survey, 27% of New Yorkers live in households earning less than $25,000 per year. Once again New York State minorities bear the burden of poverty, with 2.3% of Hispanics, 21.6% of African Americans, and 16% of Asians living below the poverty level as compared with only 8.6% of non-Hispanic Whites (New York State Minority Health Surveillance).

- According to the USDA Economic Research Service (2008), 22.5% of American children are considered to have low or very low food security. Thirty-two percent of Black non-Hispanic, 32% of Hispanic, and 15% of White U.S. households with children report food insecurity per the 2008 survey (Household Food Security 2008).

- The percentage of high school graduates in New York is 79% as compared with 80% nationwide and there are 27% college graduates verses 24% nationwide.

- The median household income in New York from 2007 is $53,448 as compared with $50,740 nationwide. Average family size in New York is 3.22 people and 53% own their homes.

- Eighty percent of New Yorkers are born in the United States while 20% are foreign born. Sixty-one percent of New Yorkers are employed in the work force and 24% utilize public transportation. The majority of New Yorkers work in management and professional occupations (U.S. Census Bureau, Census 2000).

- As of 2005, the largest racial and ethnic groups that make up the New York State population are White non-Hispanic (60%), Hispanic (16%), African American (15%), and Asian non-Hispanic (7%). The percentage of Hispanic New Yorkers increased nearly 30% between 1990 and 2000 whereas White non-Hispanics were the only group to decrease in proportion by -5.6% during the same time period (New York State Minority Health Surveillance).

60

- Ninety percent of New Yorkers are considered food secure as compared with 92% of all Americans (USDA Agriculture Research Service).

- The percentage of New York adults diagnosed with diabetes, an obesity-related disease, also varies by race with 12.3% of Black non-Hispanic, 7.5% of Hispanic, and 6.5% of White New Yorkers diagnosed (New York State Minority Health Surveillance, 2007).

- The percentages of New York high school students considered obese are 15% of Hispanics and 11% of Black non-Hispanics compared with only 9% of White non-Hispanics (New York State Minority Health Surveillance Report, 2007). Additionally, only 87% of Black non-Hispanic and 88% of Hispanic high school students reported vigorous or moderate activity during the past week using data from the Youth Risk Behavior Survey of 2005. This is significantly less as compared to the 95% of White non-Hispanic students reporting activity within the last week.

- More than half of all Black non-Hispanic and Hispanic high school students reported watching more than 3 hours per day of television according to the same survey (New York State Minority Health Surveillance Report, 2007).

- Across all racial and ethnic groups, cardiovascular disease is New York's leading killer, accounting for 43% of all deaths. In 2001, over 70% of all deaths that occurred in New York State were due to chronic diseases. Diabetes affects 1 of every 12 New Yorkers and since 1994 the number of those affected has doubled. At current rates experts suggest another doubling by the year 2050. The percentage of obese adults in New York State more than doubled from 10% in 1997 to 25% in 2008 (NYS Department of Health).

- New York State has a number of strong initiatives tackling the issue of obesity and overweight in children. The state's Strategic Plan for Overweight and Obesity Prevention outlines three priorities of focus: 1) increase the proportion of New Yorkers who are physically active; 2) increase perception of obesity as a public health risk and use of body mass index to improve early recognition; and 3) increase access to healthy food choices, particularly by low-income populations (NYS Department of Health Strategic Plan for Overweight and Obesity Prevention).

- New York State has outlined some specific goals to reduce the percentage of New York children who are overweight by the year 2013. For example, the state aims to reduce the percentage of children (ages 2-4 years) enrolled in the Supplemental Nutrition Program for Women, Infants, and Children Program (WIC) who are obese to no more than 11.6%, reduce the percentage of children ages 6-11 years who are obese to no more than 5%, and reduce the percentage of children ages 12-19 years who are obese to no more than 5%. (NYS Department of Health Priority Area: Physical Activity and Nutrition).

- With these priorities in mind the Activ8Kids! Nutrition and physical activity initiative was launched in 2005 to fight obesity and promote healthy lifestyles among children. Prior to the age of 8 years children are tasked with the daily challenge of consuming at least five fruits and vegetables, engaging in at least one hour of physical activity, and reducing screen time to fewer than two hours. This plan amounts to the 5 + 1 + 2 = 8, or Activ8 (NYS Department of Health, Activ8Kids!).

- State surveillance programs such as the Student Weight Status Initiative have been implemented to increase screening and early recognition of overweight and obesity by pediatric healthcare providers, collect and report weight status for public schools, provide local, county, and statewide estimates of the prevalence of childhood obesity, and identify successful programs to reduce childhood obesity (NYS Department of Health Priority Area: Physical Activity and Nutrition).

- An initiative targeting low-income preschool children and their families called Eat Well Play Hard (EWPH) promotes childhood obesity prevention within large-scale public health food and nutrition programs. The goals of the program include increasing developmentally appropriate physical activity, increasing consumption of low-fat dairy products, increasing consumption of fruits and vegetables, and decreasing TV and screen time. (NYS Department of Health Eat Well Play Hard).

- The New York State Strategic Alliance for Health specifically focuses on populations most in need such as select racial and ethnic groups, those limited by income and insurance coverage, and those with chronic disease. The Centers for Disease Control and Prevention has awarded a five-year grant of $2.6 million to New York City to target the South Bronx and East and Central Harlem communities to address physical activity, nutrition, and tobacco use within schools and the community. According to the NYC Department of Health and Mental Hygiene 2007 Community Health Survey, 34% of South Bronx residents and 29% of East and Central Harlem residents were obese as compared with the citywide figure of 22% (The New York Academy of Medicine NYC Strategic Alliance for Health).

- New York was selected as one of 10 states awarded with a grant from the National Governors Association to develop and disseminate model guidelines on nutrition, physical activity, and media use in after-school care settings. Using a combination of nutrition interventions in schools, worksites, and the community, the NYS Healthy Heart Program has promoted healthier behaviors while reducing disease risk associated with heart health.

- Nationwide, Michelle Obama has recently launched the Lets Move! Campaign to "solve childhood obesity within a generation." The campaign provides support to parents, promotes healthier school meals, encourages physical activity among youth, and strives to make affordable food available throughout America. The Healthier U.S. Schools Challenge established rigorous standards for school food quality, participation in meal programs, and physical activity and nutrition education. Major school food suppliers have agreed to meet the Institute of Medicine's recommendations within five years to decrease the amount of sugar, fat, and salt in school meals; increase whole grains; and double the amount of produce they serve within 10 years (Let's Move Campaign).

Environmental Characteristics

- The Centers for Disease Control and Prevention (CDC) lists contributing factors as caloric imbalance, environmental or community setting, genetics or illness, and medication use (CDC, Overweight and Obesity Causes and Consequences).

- The food industry greatly influences child food and beverage consumption, even in schools. Sugar-sweetened beverages have infiltrated our schools and market directly to the child consumer. Researchers show that these sugary beverages are a prime contributor to weight gain and obesity, contributing almost 11 percent of children's total calorie consumption. Using data between 1999 and 2004, 85% of adolescents drank an average of 30 oz. of sugar-sweetened beverages contributing an extra 356 kcal per day (Wang, Bleich & Gortmaker,

2008). In fact, each additional daily serving of soda increases a child's risk of becoming obese by 60% (Ludwig, Peterson & Gortmaker, 2001). Removing these nutrient-poor drinks from all schools would reduce exposure and access according to the Institute of Medicine (2007). As Frieden et al. (2010) point out, "more effort is needed to ensure that all food and beverages, including those available outside school meal programs, meet nutrition standards."

- Although there is no data on elementary school students, the 2007 National Youth Risk Behavior Survey indicates that 62% of high school students did not meet recommended levels of physical activity, 17% did not participate in 60 or more minutes of physical activity during the past 7 days and 35% watched 3 or more hours of television on an average school day (Healthy Kids Healthy NYS). Per the *Physical Activity Guidelines for Americans*, children and adolescents should perform 60 minutes or more of physical activity daily (*Physical Activity Guidelines for Americans*, 2008).

- Physical inactivity is a concern among New York State students as well. Sixty-two percent of high school students did not meet recommended levels of physical activity, 87% did not attend physical education classes daily, 35% watched television 3 or more hours per day on an average school day, and 29% played video or computer games or used a computer for something that was not school work for 3 or more hours per day on an average school day. (2007 National YBRSS)

- TV viewing, another contributor to increased obesity prevalence, has been shown to promote high-fat and -sugar foods aimed directly to children (Matheson, et al. 2004). Halford et al. (2004) reported that children tend to eat more of these foods after being exposed to television advertising. The impact of television viewing on obesity is most likely due to both the displacement of more vigorous activities by television and effects on diet per Dietz & Gortmaker (1985).

- Exercise helps to regulate metabolic rate, helps improve overall psychological outlook, and may aid in controlling appetite (Roberts, 2000). According to the U.S. Department of Health and Human Services, children and adolescents should do 60 minutes or more of physical activity daily (DHHS, 2008). In reality, physical activity levels have decreased by more than a third between ages nine and fifteen per the Council on Sports Medicine and Fitness (2006). Once mandatory, these structured physical education class have been omitted in many schools due to budget constraints and academic pressures (Symons, Cinelli, James & Groff, 1997).

- Brownson (2001) reports that children living in lower-SES neighborhoods have less access to safe walkways, fields, fitness clubs, and parks as compared to those living in more affluent neighborhoods, further reducing frequency of physical activity among this population.

- In New York State the top produce purchases are apples, lettuce, tomatoes, carrots, grapes, watermelon, peppers, onions, potatoes, cucumbers, bananas, and oranges. Sixty-one percent of New York schools serve meals featuring local state foods and 63% provide cooking lessons to students. While 91% of schools boast a nutrition education plan, only 36% provide education about state food and agriculture, 49% have taken students to visit a farm or farmer's market, and only 31% have planted a garden (NYS Farm to School, 2010).

- An examination of the nutritional content of the diet of New Yorkers (aged 2 years and over) shows that only 8% are consuming adequate fiber, 16% are consuming adequate vegetables, 17% are consuming recommended amounts of dairy, and 25% are consuming recommended fruit servings (USDA, Agricultural Research Service).

- Among New York State High School Students, 88% self-reported that they drank less than 3 glasses of milk per day during the past seven days and 24% self-reported that they drank a can, bottle, or glass of soda or pop at least one time a day during the past seven days (Healthy Kids, Healthy NYS). The New York State YBRSS of 2005 shows that only 21% of students ate fruits and vegetables, excluding French fries, fried potatoes, or potato chips, five or more times per day during the 7 days prior to the survey (YRBSS, 2005).

Socioeconomic Characteristics

- Obesity is particularly common among Hispanic, African American, Native American, and Pacific Islander women (*Healthy People 2010*).

- Hispanics and Native American children of both genders along with African American girls share the highest rates of overweight in the nation. Second and third generation children of Pacific Islander and Asian American families are becoming more susceptible (Crawford, 2001).

- According to data from the Behavioral Risk Factor Surveillance System surveys, United States, 2006–2008, non-Hispanic blacks had a 51% greater prevalence of obesity, and Hispanics had a 21% greater prevalence, when compared with non-Hispanic whites.

- During 2006–2008, the age-adjusted estimated prevalence of obesity was 35.7% among Non-Hispanic blacks, 28.7% among Hispanics, and 23.7% among non-Hispanic whites (CDC Morbidity and Mortality Weekly Report, 2009).

- Medical expenditures for obese workers, depending on severity of obesity and sex, are 29% to 117% greater than expenditures for workers of average weight (CDC Overweight and Obesity Economic Consequences).

- The economic cost of obesity in the United States was about $117 billion in 2000 (Office of the Surgeon General).

- Between 1987 and 2001, disease associated with obesity accounted for 27% of the increases in medical costs and by 2000, obesity-related health care costs totaled $117 billion. According to NYS Comptroller DiNapoli, New York ranks second among states in adult obesity-related medical expenditures, with total spending in New York increasing to $7.6 billion, 81% of which is paid by Medicaid and Medicare (NYS Department of Health Obesity Prevention).

- In 2003, New York ranked second highest among all states in total adult obesity-related medical expenditures, with estimated spending of nearly $6.1 billion (New York State Health Eating and Physical Activity Alliance).

Target Population Data

- Childhood overweight and obesity has increased dramatically over the past few decades. In fact, the prevalence of overweight among 6- to 11-year-old American children has ballooned from approximately 6.5% in 1976 to more than 19.6% by 2007-2008 (Ogden et al., 2010; Ogden, Carroll and Flegal, 2008).

- Obesity rates in New York mirror those of the nation with 20% of elementary school children in New York State obese and 24% of New York City elementary school children obese (Health Kids, Healthy NYS). Sixteen percent of New York youth grades 9 to 12 are overweight, and another 11% are obese, according to 2007 Youth Risk Behavior Survey data. Only 38% of youth in these grade levels are meeting current physical activity recommendation levels. Approximately one quarter drink at least one non-diet soda each day. Thirty-five percent watch more than three hours of television each day (Youth Risk Behavior Survey, 2007). Of the 4.5 million New Yorkers under the age of 18, estimates suggest that 1.1 million are obese, or one in four young people (Office of the State Comptroller, 2008). This translates to an estimated annual $242 million in medical costs attributed to these children of New York State (Office of the State Comptroller, 2008).

- For children and adolescents (aged 2–19 years), the body mass index (BMI) value is plotted on the CDC growth charts to determine the corresponding BMI-for-age percentile. Overweight is defined as a BMI at or above the 85th percentile and lower than the 95th percentile while obesity is defined as a BMI at or above the 95th percentile for children of the same age and sex (Center for Disease Control and Prevention).

- Twenty-three percent of Mexican-American and 21% of Non-Hispanic black adolescents ages 12-19 are more likely to be overweight than the 14% of overweight non-Hispanic white adolescents. Mexican-American children ages 6-11 were more likely to be overweight (22%) than non-Hispanic black children (20%) and non-Hispanic white children (14%) (National Center for Health Statistics).

- In children 6-11 years old, 22% of Mexican American children were overweight, whereas 20% of African American children and 14% of non-Hispanic White children were overweight. In addition to the children and teens who were overweight in 1999-2002, another 15% were at risk of becoming overweight (National Center for Health Statistics, 2004).

- In a national survey of American Indian children 5-18 years old, 39% were found to be overweight or at risk for overweight (Jackson, 1993).

- Using the BRFSS 2006-2008 data, New York State showed 29.7% of Black non-Hispanics, 27.1% of Hispanics, and 22.8% of White non-Hispanics as obese (CDC Morbidity and Mortality Weekly Report, 2009).

- The proportion of obese children in New York State is between 16% and 31% and those from a minority background share a greater burden of the disease. For example, the obesity rate for White non-Hispanic elementary school children in New York State is 18% compared to 23% of African Americans and 30% of Hispanics classified as obese (Upstate NY, Grade 3 Oral Health, Physical Activity, and Nutrition Survey). These rates greatly exceed the U.S. NHANES 1999-2002 survey rate of 16% and the *Healthy People 2010* target of 5% (*Healthy People 2010*).

- Mexican American boys tend to have a higher prevalence of overweight than non-Hispanic black or non-Hispanic white boys. Non-Hispanic black girls tend to have a higher prevalence of overweight than Mexican American or non-Hispanic white girls. Non-Hispanic white adolescents from lower-income families experience a greater prevalence of overweight than those from higher-income families (Office of the Surgeon General).

Executive Summary

1. Obesity prevalence among children nationwide and in New York State is increasing and those most effected are children of minority populations:
Childhood overweight and obesity have increased dramatically over the past few decades. In fact, the prevalence of overweight among 6- to 11-year-old American children has ballooned from approximately 6.5% in 1976 to more than 19.6% by 2007-2008 (Ogden et al., 2010; Ogden, Carroll and Flegal, 2008). *Healthy People 2010* named overweight and obesity as one of the leading health indicators and has included an objective to reduce overweight and obesity in children and adolescents as part of *Healthy People 2020* (*Healthy People 2010*).

Of the 4.5 million New Yorkers under the age of 18, estimates suggest that 1.1 million are obese, or one in four young people (Office of the State Comptroller, 2008). Obesity rates in New York mirror those of the nation with 20% of elementary school children in New York State obese and 24% of New York City elementary school children obese (Health Kids, Healthy NYS). Sixteen percent of New York youth grades 9 to 12 are overweight, and another 11% are obese, according to 2007 Youth Risk Behavior Survey data. Adding to the problem is the concern that only 38% of youth in these grade levels are meeting current physical activity recommendation levels. Approximately one quarter drink at least one non-diet soda each day and 35% watch more than three hours of television each day (Youth Risk Behavior Survey, 2007).

The prevalence of childhood obesity among children of minority populations is particularly alarming. Hispanics and Native American children of both genders along with African American girls share the highest rates of overweight in the nation. Second and third generation children of Pacific Islander and Asian American families are becoming more susceptible (Crawford, 2001). According to data from the Behavioral Risk Factor Surveillance System surveys, United States, 2006–2008, non-Hispanic blacks had a 51% greater prevalence of obesity, and Hispanics had a 21% greater prevalence, when compared with non-Hispanic Whites.

This pattern was consistent across most U.S. states. During 2006–2008, the age-adjusted estimated prevalence of obesity was 35.7% among Non-Hispanic Blacks, 28.7% among Hispanics, and 23.7% among non-Hispanic Whites (CDC Morbidity and Mortality Weekly Report, 2009). Twenty-three percent of Mexican-American and 21% of Non-Hispanic black adolescents ages 12-19 were overweight as compared with only 14% of non-Hispanic White adolescents. In children 6-11 years old, 22% of Mexican American children were overweight, whereas 20% of African American children and 14% of non-Hispanic White children were overweight (National Center for Health Statistics). In a national survey of American Indian children 5-18 years old, 39% were found to be overweight or at risk for overweight (Jackson, 1993). In addition to the children and teens who were overweight in 1999-2002, another 15% were at risk of becoming overweight (National Center for Health Statistics, 2004).

Using the BRFSS 2006-2008 data, New York State showed 29.7% of Black non-Hispanics, 27.1% of Hispanics, and 22.8% of White non-Hispanic as obese (CDC Morbidity and Mortality Weekly Report, 2009). Among elementary school children in New York State, the obesity rate for White non-Hispanics was 18% compared to 23% for African Americans and 30% for Hispanics (Upstate NY, Grade 3 Oral Health, Physical Activity, and Nutrition Survey). These rates greatly exceed the U.S. NHANES 1999-2002 survey rate of 16% and the *Healthy People 2010* target of 5% (*Healthy People 2010*). The percentages of New York high school students considered obese are 15% of Hispanics and 11% of Black non-Hispanics as compared with only 9% of White non-Hispanics (New York State Minority Health Surveillance Report, 2007). Additionally, only 87% of Black non-Hispanic and 88% of Hispanic high school students reported vigorous or moderate activity during the past week using data from the Youth Risk Behavior Survey of 2005. This is significantly less as compared to the 95% of White non-Hispanic students reporting activity within the last week. More than half of all Black non-Hispanic and Hispanic high school students reported watching more than 3 hours per day of television according to the same survey (New York State Minority Health Surveillance Report, 2007).

2. Childhood obesity places children at greater risk for obesity-related disease later in life: Approximately one-half of overweight school-aged children remain overweight as adults (Serdula, 1993). Whitaker et al. found that approximately 80% of children who were overweight at age 10-15 years were obese adults at age 25 years (Whitaker, 1997). Authors Lauer et al. found that obesity acquired in childhood is predictive of adult obesity and development of coronary artery calcification (Lauer, Clarke, & Burns, 1997). Additional health problems such as high rates of hypertension, asthma, musculoskeletal discomfort, shortness of breath, and obstructive sleep apnea have all been reported in overweight children (Waters, & Baur, 2003). Obese children also suffer from low self-esteem and significantly higher rates of sadness, loneliness, and nervousness, and are more likely to engage in high-risk behaviors such as smoking or consuming alcohol (Strauss, 2000). Therefore the key strategy recognized by most professionals for controlling the obesity epidemic is prevention (Muller & Mast, 2001).

3. Factors contributing to obesity: There are many factors contributing to the complexity of the national obesity epidemic. The Centers for Disease Control and Prevention (CDC) lists contributing factors as caloric imbalance, environmental or community setting, genetics or illness, and medication use (CDC, Overweight and Obesity Causes and Consequences).

The American diet has changed dramatically over the past few decades with the introduction of highly processed foods, low in nutrient density and high in fat, calories, and sugar. The food industry greatly influences child food and beverage consumption, even in schools. Sugar-sweetened beverages have infiltrated our schools and market directly to the child consumer. Researchers show that these sugary beverages are a prime contributor to weight gain and obesity, contributing almost 11% of children's total calorie consumption. Using data from between 1999 and 2004, 85% of adolescents drank an average of 30 oz. of sugar-sweetened beverages contributing an extra 356 kcal per day (Wang, Bleich & Gortmaker 2008). In fact, each additional daily serving of soda increases a child's risk of becoming obese by 60% (Ludwig, Peterson & Gortmaker, 2001). According to the Institute of Medicine, removing these nutrient-poor drinks from all schools would reduce exposure and access (2007). As Frieden et al. (2010) point out, "More effort is needed to ensure that all food and beverages, including those available outside school meal programs, meet nutrition standards."

An examination of the nutritional content of the diet of New Yorkers (aged 2 years and over) shows that only 8% are consuming adequate fiber, 16% are consuming adequate vegetables, 17% are consuming recommended amounts of dairy, and 25% are consuming recommended fruit servings (USDA, Agriculture Research Service). Among New York State High School Students,

88% self-reported that they drank less than 3 glasses of milk per day during the past seven days and 24% self-reported that they drank a can, bottle, or glass of soda or pop at least one time a day during the past seven days (Healthy Kids, Healthy NYS). The New York State YBRSS of 2005 shows that only 21% of students ate fruits and vegetables, excluding French fries, fried potatoes, or potato chips, five or more times per day during the 7 days prior to the survey (YRBSS, 2005). While 91% of New York schools boast a nutrition education plan only 36% provide education about state food and agriculture, 49% have taken students to visit a farm or farmer's market, and only 31% have planted a garden (NYS Farm to School, 2010).

Physical inactivity is a concern among New York State students as well. Sixty-two percent of high school students did not meet recommended levels of physical activity, 87% did not attend physical education classes daily, 35% watched television 3 or more hours per day on an average school day, and 29% played video or computer games or used a computer for something that was not school work for 3 or more hours per day on an average school day (2007 National YBRSS).

Interventions to target minority youth are important to slow the obesity rends in children: New York State has a number of strong initiatives tackling the issue of obesity and overweight in children. The state's Strategic Plan for Overweight and Obesity Prevention outlines three priorities of focus: 1) increase the proportion of New Yorkers who are physically active; 2) increase perception of obesity as a public health risk and use of body mass index to improve early recognition; and 3) increase access to healthy food choices, particularly by low-income populations (NYS Department of Health Strategic Plan for Overweight and Obesity Prevention). With these priorities in mind the Activ8Kids! Nutrition and physical activity initiative was launched in 2005 to fight obesity and promote healthy lifestyles among children. Prior to the age of 8 years children are tasked with the daily challenge of consuming at least five fruits and vegetables, engaging in at least one hour of physical activity, and reducing screen time to fewer than two hours. This plan amounts to the 5 + 1 + 2 = 8, or Activ8 (NYS Department of Health, Activ8Kids!).

The New York State Strategic Alliance for Health specifically focuses on populations most in need such as select racial and ethnic groups, those limited by income and insurance coverage, and those with chronic disease. The Centers for Disease Control and Prevention has awarded a five-year grant of $2.6 million to New York City to target the South Bronx and East and Central Harlem communities to address physical activity, nutrition, and tobacco use within schools and the community. According to the NYC Department of Health and Mental Hygiene 2007 Community Health Survey, 34% of South Bronx residents and 29% of East and Central Harlem residents were obese as compared with the citywide figure of 22% (The New York Academy of Medicine NYC Strategic Alliance for Health).

Feedback

Community Parent:
The data supports the need for an intervention among our children, yet how are parents supposed to get involved? School food is not meeting our child's nutritional needs and purchasing healthy foods is costly and they are sometimes unavailable in my area.

Will there be more effort to bring healthy and affordable foods to my community?

68

Pediatrician

The BMI surveillance system requires trained staff and equipment. Who will be providing the equipment and training required to maintain BMI records of all pediatric patients?

References

Brownson, R. (2001). Environmental and policy determinants of physical activity in the U.S. *American Journal of Public Health*, 91, 1995–2003.

Center for Disease Control. Healthy Weight Body Mass Index. Retrieved on June 2, 2010 from http://www.cdc.gov/healthyweight/assessing/bmi/

Center for Disease Control. Morbidity and Mortality Weekly Report (MMWR) Differences in Prevalence of Obesity Among Black, White, and Hispanic Adults --- United States, 2006—2008 July 17, 2009 / 58(27);740-744

http://www.cdc.gov/mmwr/preview/mmwrhtml/mm5827a2.htm

Center for Disease Control. Overweight and Obesity Economic Consequences. Retrieved on May 28, 2010 from http://www.cdc.gov/obesity/causes/economics.html

http://www.surgeongeneral.gov/topics/obesity/calltoaction/fact_glance.html

Council on Sports Medicine and Fitness Council on School Health. (2006). Active healthy living: prevention of childhood obesity through increased physical activity. *Pediatrics*, 117(5), 1834–42.

Crawford PB, Story M, Wang MC, Ritchie LD, Sabry ZI. Ethnic issues in the epidemiology of childhood obesity. *Ped Clin N Am* 2001;48:855-78.

Department of Health and Human Services. (2008). Physical activity guidelines for Americans 2008. Retrieved on April 2, 2010 from: http://www.health.gov/

deSilva-Sanigorski A.M., Bell A.C., Kremer P., Nichols M., Crellin M., Smith M., et al. (2010). Reducing obesity in early childhood: results from Romp & Chomp, an Australian community-wide intervention program. *American Journal of Clinical Nutrition*, 91(4), 831-40.

Dietz W.H., Gortmaker S.L. (1985). Do we fatten our children at the TV set? Obesity and Television viewing in children and adolescents. *Pediatrics*, 75, 807-812.

Finkelstein E.A., Trogdon J.G., Cohen J.W., Dietz W. (2009). Annual medical spending attributable to obesity: payer- and service-specific estimates. *Health Affairs*, 28(5), 822–31.

Frieden, T.R., Dietz, W., (2010). Reducing childhood obesity through policy change: acting now to prevent obesity. *Health Affairs*, 29(3), 357-63.

Halford, J.C., Boyland, E.J., Hughes, G., Oliveira, L.P., Dovey, T.M. (2004). Effect of television advertisements for food consumption in children. *Appetite*, 42(2), 221-225.

Health Kids, Healthy New York. After-School Initiative Toolkit. (2008). New York State Healthy Easting and Physical Activity Alliance. October 27.

Institute of Medicine (2007). *Nutrition standards for foods in schools: leading the way toward healthier youth*. Washington (DC): National Academies Press.

Jackson, Yvonne. (1993) Height, weight, and body mass index of American Indian schoolchildren, 1990-1991. *Journal of the American Dietetic Association*. 93(10) 1136-1140.

Lauer, R.M., Clarke, W.R. & Burns, T.L. (1997). Obesity in childhood: the Muscatine Study. *Acta Paeditrica*, 38(6), 432–437.

Let's Move Campaign. (Last modified April 2, 2010). Let's Move: American's Move to Raise a Healthier Generation of Kids. Retrieved on April 2, 2010 from: http://www.letsmove.gov/

Ludwig D.S., Peterson K.E., Gortmaker S.L. (2001). Relation between consumption of sugar-sweetened drinks and childhood obesity: a prospective, observational analysis. *Lancet*, 357(9255), 505–8.

Matheson, D., et al. (2004). Children's food consumption during television viewing. American *Journal of Clinical Nutrition*, 79 (6), 1088–1094.

Muller, M.J., Mast, M., Asbeck, I., Langnase, K., Grund, A. (2001). Prevention of obesity--is it possible? *Obesity Review*, 2, 15-28.

National Center for Health Statistics. (2004). *Obesity Still a Major Problem, New Data Show*. October 6. Retrieved on May 28, 2010 from http://www.cdc.gov/nchs/pressroom/04facts/obesity.htm

National Center for Health Statistics. *Prevalence of Overweight Among Children and Adolescents: United States, 1999-2002*. Retrieved on June 2, 2010 from

http://aspe.hhs.gov/health/reports/child_obesity/#_ftn4

The New York Academy of Medicine. Health Policy NYC Strategic Alliance for Health. Retrieved on May 28, 2010 from

http://www.nyam.org/initiatives/sp-safh7.shtml

New York State Department of Agriculture & Markets. (2010). New York State Farm to School. Farm to School Coordinating Committee Meeting. March 23.

New York State Department of Health. *Activ8Kids!* Healthy Lifestyles & Prevention. Retrieved on May 28, 2010 from http://www.health.state.ny.us/prevention/obesity/activ8kids/index.htm

New York State Department of Health. *Eat Well Play Hard*. Retrieved on May 28, 2010 from

http://www.health.state.ny.us/prevention/nutrition/resources/eat_well_play_hard/

New York State Department of Health. Obesity Prevention. *Obesity Statistics and the Impact of Obesity*. Retrieved on May 28, 2010 from

http://www.health.state.ny.us/prevention/obesity/statistics_and_impact/

New York State Department of Health. *Physical Activity and Nutrition*. Retrieved on May 28, 2010 from

http://www.health.state.ny.us/prevention/prevention_agenda/physical_activity_and_nutrition/index.htm

New York State Department of Health. *Population Survey: Percent of Population Below 100% and 200% of the Federal Poverty Level*. Retrieved on May 28, 2010 from http://www.health.state.ny.us/statistics/chac/general/poverty_level.htm

New York State Department of Health. *Strategic Plan for Overweight and Obesity Prevention.* Executive Summary. Retrieved on June 2, 2010 from http://www.health.state.ny.us/prevention/obesity/strategic_plan/executive_summ.htm

New York State Healthy Eating and Physical Activity Alliance. *New York's Children Suffer From Preventable Health Crisis.* Retrieved on June 16, 2010 from

http://nyshepa.org/news/?p=113

New York State Minority Health Surveillance. (2007). New York State Department of Health.

Office of the State Comptroller. (2008). *Preventing and Reducing Childhood Obesity in New York.*

Office of the Surgeon General. Department of Health and Human Services. Retrieved on June 16, 2010 from

http://www.surgeongeneral.gov/topics/obesity/calltoaction/fact_glance.html

Ogden CL, Carroll MD, Curtin LR, Lamb MM, Flegal KM. (2010). Prevalence of High Body Mass Index in US Children and Adolescents, 2007–2008. *JAMA.* 303:242–249

Ogden CL, Carroll MD, Flegal KM. (2008). High Body Mass Index for Age Among US Children and Adolescents, 2003–2006. *JAMA.* 299:2401–2405.

Roberts, S. (2000). The role of physical activity in the prevention and treatment

of childhood obesity. *Pediatric Nursing,* 26(1), 33–41.

Serdula MK, Ivery D, Coates RJ, Freedman DS, Williamson DF, Byers T. (1993). Do obese children become obese adults? A review of the literature *Prev Med* 22:167-77.

Strauss, R. (2000). Childhood obesity and self-esteem. *Pediatrics,* 105(1), 15–16.

Symons C.W., Cinelli B., James T.C. & Groff, P. (1997). Bridging student health risks and academic achievement through comprehensive school health programs. *Journal of School Health,* 67(6), 220–7.

Upstate NY, Grade 3 Oral Health. (2004). Physical Activity, and Nutrition Survey, 2004. New York City; *Am J Public Health,* 94: 1498.

U.S. Census Bureau. State and County QuickFacts. Consolidated Federal Funds Report Last Revised: Thursday, 22-Apr-2010 08:35:19 EDT. Retrieved from

http://quickfacts.census.gov/qfd/states/36000.html

U.S. Census Bureau, Census 2000. Retrieved on June 10, 2010 from http://www.census.gov/main/www/cen2000.html

U.S. Department of Agriculture. Agriculture Research Service. The Community Nutrition Mapping Project. Retrieved on May 28, 2010 from

http://www.ars.usda.gov/Services/docs.htm?docid=15713

U.S. Department of Agriculture. *Healthy People 2010.* Nutrition and Overweight Chapter 19. Retrieved on May 28, 2010 from

http://www.healthypeople.gov/Document/HTML/Volume2/19Nutrition.htm#_Toc490383123

U.S. Department of Health and Human Services. Office of Disease Prevention and Health Promotion. *Healthy People 2010*. (2010). Chapter 19 Nutrition and Overweight. Retrieved on May 28, 2010 from

http://www.healthypeople.gov/Document/HTML/Volume2/19Nutrition.htm#_Toc490383123

Wang, Y.C., Bleich, S.N., Gortmaker, S.L. (2008). Increasing caloric contribution from sugar-sweetened beverages and 100 percent fruit juices among U.S. children and adolescents, 1988–2004. *Pediatrics*, 121(6), e1604–14.

Waters, E. & Baur, L. (2003). Childhood obesity: modernity's scourge. *Medical Journal of Australia*, 178 (9), 422–423.

Whitaker RC, Wright JA, Pepe MS, Seidel KD, Dietz WH. Predicting obesity in young adulthood from childhood and parental obesity. *N Engl J Med* 1997; 37(13):869–873.

Youth Risk Behavior Survey. (2005). Center for Disease Control. Retrieved on June 2, 2010 from http://www.cdc.gov/HealthyYouth/yrbs/index.htm

CPSIA information can be obtained
at www.ICGtesting.com
Printed in the USA
FFOW030934131212
531FF